Computational Intelligence for Connective Cognition Networks

This book explores deep learning architectures such as convolutional neural networks and recurrent neural networks for tasks like image analysis, speech recognition, and natural language processing within network paradigms. It uses machine learning algorithms such as neural networks, support vector machines, and decision trees for data analysis and prediction tasks.

This book:

- Covers a wide range of topics within network paradigms, including intelligence modeling, sustainability, quantum computing, and network security.
- Utilizes various machine learning algorithms such as neural networks, support vector machines, and decision trees for data analysis, and prediction tasks.
- Addresses contemporary issues like fake news detection, social media analysis, and cybersecurity.
- Employs network analysis techniques to understand the structure and dynamics of complex systems, including social networks, communication networks, and biological networks.
- Explores the integration of quantum computing principles and algorithms to solve computational intelligence tasks efficiently, especially in quantum-based network paradigms.

It is primarily written for senior undergraduates, graduate students, and academic researchers in the fields including electrical engineering, electronics and communications engineering, computer engineering, and information technology.

Advancing Science and Engineering through Artificial Intelligence, Machine Learning, and Mathematical Modeling

Innovation in Healthtech: A Roadmap for Empowering Healthcare
Edited by: Rakesh Kumar and Meenu Gupta

Computational Intelligence for Connective Cognition Networks: Advances and Applications
Edited by: Kirti Aggarwal, Anuja Arora, Zahid Akhtar, and Alessandro Bruno

Computational Intelligence for Connective Cognition Networks

Advances and Applications

Edited by
Kirti Aggarwal, Anuja Arora, Zahid Akhtar,
and Alessandro Bruno

CRC Press
Taylor & Francis Group
Boca Raton London New York

CRC Press is an imprint of the
Taylor & Francis Group, an **informa** business

First edition published 2026
by CRC Press
2385 NW Executive Center Drive, Suite 320, Boca Raton FL 33431

and by CRC Press
4 Park Square, Milton Park, Abingdon, Oxon, OX14 4RN

CRC Press is an imprint of Taylor & Francis Group, LLC

ISBN: 9781032942346 (hbk)
ISBN: 9781032942353 (pbk)
ISBN: 9781003569619 (ebk)

DOI: 10.1201/9781003569619

Typeset in Sabon
by Newgen Publishing UK

Contents

Preface

Hey there, and welcome to the first edition of *Computational Intelligence for Connective Cognition Networks: Advances and Applications*! This book feels like it's arriving at just the right time—a moment when computational intelligence is hitting its stride, blending clever algorithms, sprawling networks, and ideas that actually matter to people's lives. It's changing how we see the world and how we tackle its messier problems. To me, "connective cognition networks" is all about that sweet spot where machines start thinking a bit more like us and networks tie it all together, letting us take on challenges across all sorts of fields.

What you'll find in these pages is a mix of fresh ideas and real-world grit. We've got chapters that dig into everything from sharpening ECG analysis with some fancy modeling and machine learning tricks to poking around the wild possibilities of quantum networking. It's a broad sweep—think genetic algorithms sprucing up images, security setups for drones, or figuring out what's fake in photos and videos. Each piece here nudges the limits of what smart tech can do. And it doesn't stop there—we're also looking at how AI can make roads safer, sort out multilingual documents, sniff out fake news, and even help teachers prep better lessons with big language models. It's all about the potential to shake things up.

This book didn't just happen on its own—it's the work of an amazing crew of researchers, hands-on experts, and big thinkers who've poured their know-how into it. They've not only pushed the theory behind connective cognition networks further but also shown how it can solve real problems out there in the world. As editors, we've tried to pull all these threads into something that makes sense together—something that's deep but not intimidating, so whether you're a pro in this stuff or just curious, there's something here for you.

Putting this together has been quite a ride, full of "aha" moments and back-and-forth chats. I can't thank the authors enough for bringing their A-game, or the reviewers for catching things we missed and making it all sharper. And a big shoutout to the publishing folks who stuck with us to get this out the door.

We hope Connective Cognition Networks: Advances and Applications lights a spark—maybe it'll teach you something new, get you thinking, or even push you to question where all this tech is headed. Whether you're a researcher, a pro in the field, or a student dipping your toes in, we want you to see what's happening now and maybe catch a peek at what's coming next. These tools and ideas are weaving new links between data, machines, and how we think—opening doors we're only starting to peek through.

So, dive in, flip through, and come along with us on this journey into connective cognition networks. We're excited to have you here!

Acknowledgments

Pulling together *Computational Intelligence for Connective Cognition Networks: Advances and Applications* has been a team effort, and I'm genuinely grateful for everyone who's had a hand in it. This book wouldn't be what it is without a whole bunch of folks who've chipped in with their time, talent, and encouragement.

First off, a massive thank you to all the authors who've poured their hearts and minds into these chapters. You've brought such a range of ideas and expertise to the table—everything from ECG analysis to quantum networking and beyond. It's been a real privilege to see your work come together and shape this collection. Your willingness to share, tweak, and polish has made this something special.

I also want to give a big nod to our reviewers. You've been the unsung heroes here, diving into drafts with sharp eyes and thoughtful feedback. Your suggestions didn't just catch the little stuff—they helped us sharpen the big picture, too. Thanks for keeping us on our toes.

Finally, a shoutout to the broader research community and everyone out there pushing the boundaries of computational intelligence. You're the reason a book like this even makes sense—your curiosity and drive light the way for projects like ours And to my family and friends—thanks for putting up with me disappearing into this project. Your support, whether it was a kind word or just letting me ramble about networks and AI, kept me going.

This has been a wild ride, and I'm so thankful to everyone who's ridden along with us. Here's to you all—cheers for making *Connective Cognition Networks* happen!

About the Editors

Kirti Aggarwal is working as a senior assistant professor in the Department of Computer Science & Engineering, Jaypee Institute of Information Technology, Noida, India. She received her Ph.D. degree in Computer Science & Engineering from Jaypee Institute of Technology, Noida in September 2023. She has more than 10 years of academic experience. Her research interests include social networking, nature inspired computing, optimization algorithms, DBMS, discrete mathematics, computer networks, compiler design, and java programming. She has published papers in International Journal and Conferences.

Anuja Arora is working as a professor in the Computer Science & Engineering Department of Jaypee Institute of Information Technology, Noida, India. She has academic and research experience of 19 years and industry experience of 1.5 years. She received her Ph.D. degree in Computer Science from Apaji Institute of Mathematics & Applied Computer Technology, Banasthali University, Banasthali, India in December 2013. She is a Senior IEEE Member, ACM Member, SIAM Member, INSTICC, and Life Member of IAENG. She has published more than 100 research papers in peer-reviewed international journals, book chapters, and conferences. Three students have been awarded Ph.D. under her supervision and three are in process. Her research interest includes deep learning, artificial neural network, social network analysis and mining, sustainable computing, data science, machine learning, data mining, web intelligence, web application development and web technologies, software engineering, software testing, and information retrieval systems. Dr. Arora participated in many international conferences as an organizer, session chair, and member of national advisory and International Program Committees. She is an editorial board member of numerous IGI, Inderscience, and Bentham international journals. She is the reviewer of many reputed and peer-reviewed IEEE transactions—TKDE, TNSM, IEEE Transaction of Cybernetics, etc. She is also the reviewer of various international Journals.

Zahid Akhtar received his Ph.D. in electronic and computer engineering from the University of Cagliari, Italy. He is currently an assistant professor at the Department of Network and Computer Security, State University of New York (SUNY) Polytechnic Institute, USA. Prior to that, he was a research assistant professor with the University of Memphis, USA, and a postdoctoral fellow with the INRSEMT, University of Quebec, Canada, the University of Udine, Italy, Bahcesehir University, Turkey, and the University of Cagliari. His research interests include computer vision and machine learning with applications to cybersecurity, biometrics, affect recognition, image and video processing, and audiovisual multimedia quality assessment.

Alessandro Bruno earned a Ph.D. degree in computer engineering from DINFO, Palermo University. He is now a tenure-track assistant professor with the Department of Business, Law, Economics, and Consumer Behaviour at IULM University, in Milan, Italy. Before joining IULM University, Alessandro was an assistant professor at Humanities University (Milano, Italy). He worked as a lecturer in computing with the Department of Computing and Informatics, Bournemouth University in 2021 and 2022. He covered postdoc positions at NCCA (National Centre for Computer Animation) in the UK, INAF IASF (Italian National Institute for Astrophysics), University of Palermo, and IZS (Istituto Zooprofilattico Sperimentale della Sicilia). He was a research visitor at Mullard Space Science Laboratory (MSSL), University College London (UCL) in the imaging group led by Professor Jan-Peter Muller. He is the author of more than 50 International articles and serves as a associate editor for International journals. Alessandro is currently the principal investigator of a project funded by the NGI-Search consortium named HeReFaNMi (Health-Related Fake News Mitigation). His research interests include computer vision, artificial intelligence, and image analysis. He has mostly dealt with visual attention and visual saliency, biomedical imaging, crowd behavior analysis, image and video forensics, remote sensing, and human–computer interaction.

Contributors

Apeksha Aggarwal
Indian Insitute of Technology
Roorkee, India

Kirti Aggarwal
Jaypee Institute of Information Technology
Noida, India

Preety Aneja
DAV College
Guru Nanak Dev University
Punjab, India

Anuja Arora
Jaypee Institute of Information Technology
Noida, India

Ruby Beniwal
Jaypee Institute of Information Technology
Noida, India

Simran Bhardwaj
Jaypee Institute of Information Technology
Noida, India

Nitesh Bharot
Data Science Institute
University of Galway
Galway, Ireland

Surekha Bhusnur
Bhilai Institute of Technology
Durg Chhattisgarh India

Harshita Chandwani
Jaypee Institute of Information Technology
Noida, India

Mukta Goyal
Jaypee Institute of Information Technology
Noida, India

Meenal Jain
Jaypee Institute of Information Technology
Noida, India

Shruti Kalra
Jaypee Institute of Information Technology
Noida, India

Kartikey
Jaypee Institute of Information Technology
Noida, India

Amanpreet Kaur
Jaypee Institute of Information Technology
Noida, India

Pawan Kumar
Indian Institute of Technology
Bombay, India

Sangeeta Lal
School of Computer Science & Mathematics
Keele University
England

Alka Mishra
Bhilai Institute of Technology
Durg Chhattisgarh India

Santosh Kumar Mishra
Bhilai Institute of Technology
Durg Chhattisgarh India

Monika
Jaypee Institute of Information Technology
Noida, India

R. K. Pateriya
Maulana Azad National Institute of Technology
Bhopal, India

Nikhil Rathoure
Maulana Azad National Institute of Technology
Bhopal, India

Neetu Sardana
Jaypee Institute of Information Technology
Noida, India

Pushpendra Singh
Bhilai Institute of Technology
Durg, Chhattisgarh, India

Adwitiya Sinha
TERI School of Advanced Studies
New Delhi, India

Deepak Singh Tomar
Maulana Azad National Institute of Technology
Bhopal, India

Deepika Varshney
Jaypee Institute of Information Technology
Noida, India

Priyanka Verma
University of Limerick
Ireland

Ankit Vidyarthi
Jaypee Institute of Information Technology
Noida, India

Yun Hwa Wong (Jenny)
School of Computer Science & Mathematics
Keele University
England

Chapter 1

Enhancing ECG analysis through parametric quartic spline modeling and machine learning classification

Alka Mishra, Surekha Bhusnur, Santosh Kumar Mishra, and Pushpendra Singh

1.1 INTRODUCTION

Electrocardiogram (ECG) signals play a vital role in diagnosing various cardiac abnormalities and monitoring heart health. These signals, which represent the electrical activity of the heart, provide valuable insights into the heart's functioning and can aid in the detection of conditions such as arrhythmias, ischemia, and myocardial infarction [1, 2]. As such, accurate analysis and interpretation of ECG signals are paramount in clinical practice.

Traditionally, ECG signals are analyzed visually by trained medical professionals to identify abnormalities. However, this process can be time-consuming and subjective, leading to variability in diagnosis [3]. To overcome these limitations, there has been a growing interest in the development of automated methods for ECG analysis, leveraging advancements in signal processing and machine learning techniques. One area of focus in ECG research is the development of accurate models that can replicate the complex morphology of ECG signals [4]. These models are essential for various applications, including algorithm development, medical device testing, and education.

In recent years, significant progress has been made in developing methods for generating ECG signals. Many of these methods are based on trigonometric, Fourier, and Gaussian techniques. While these methods offer simplicity and computational efficiency, they often fall short in accurately capturing the complexity and variability of real ECG data and representing certain physiological features [5], which limits their applicability in contexts requiring high fidelity and realism [6, 7]. Some approaches using neural networks for ECG modeling have also been proposed. These models excel at learning complex patterns and capturing intricate relationships in ECG data. However, they come with drawbacks such as high complexity and computational requirements, lack of interpretability, the need for large datasets, and susceptibility to overfitting [8–11].

Studies have also explored the parametric representation of ECG signals. Dolinsky et al. introduced a model based on elementary trigonometric

DOI: 10.1201/9781003569619-1

and linear functions, which was further implemented by Gerasimov and Pedonova [12, 13]. In contrast, Mishra et al. focused on the critical need for precise ECG signal analysis and classification, emphasizing the role of ECG in diagnosing cardiac electrical abnormalities. They highlighted the challenges posed by noise and artifacts in actual ECG recordings, which complicate performance assessment [14]. Parametric spline-based methods have emerged as promising approaches for ECG modeling, offering the flexibility to capture the intricate details of ECG waveforms with a minimal number of control points. In addition to ECG modeling, there is a need for robust classification algorithms capable of accurately identifying different cardiac rhythms and abnormalities from ECG signals [15–17].

ECG classification using machine learning involves the automated categorization of ECG signals into various classes based on their waveform characteristics. This process begins with the acquisition of high-quality ECG data, followed by preprocessing to remove noise and artifacts. Relevant features are then extracted from the signals, capturing important waveform characteristics such as temporal intervals and morphological features [18–21]. Machine learning models, including support vector machines, neural networks, and decision trees, are selected and trained using labelled ECG data to classify signals into different categories, such as normal and abnormal rhythms. Model performance is evaluated using metrics like accuracy and sensitivity, and optimization techniques are applied to improve classification accuracy. Once trained and validated, the optimized model can be deployed in clinical settings to assist healthcare professionals in diagnosing cardiac conditions and providing timely interventions [22, 23].

1.2 METHODOLOGY

A parametric quartic spline is a mathematical curve-fitting technique used to interpolate data points in a smooth and continuous manner. In the context of ECG modeling, it involves creating a mathematical representation of the ECG waveform based on a set of control points [24]. Here's how it works:

1. **Control Points Selection**: Control points are strategically placed along the ECG waveform to capture important morphological features such as peaks, valleys, and inflection points. The number and placement of control points are crucial as they determine the accuracy of the spline curve in representing the original ECG signal.
2. **Quartic Polynomial Interpolation**: A quartic polynomial is used to interpolate between adjacent control points. This generalized polynomial function is of the form:

$$f(x) = ax4 + bx3 + cx2 + dx + e \tag{1.1}$$

where in (1.1) a, b, c, d and e are coefficients determined by the interpolation process.

3. **Curve Smoothing**: The spline curve is smoothed to ensure continuity and minimize oscillations. This is achieved by adjusting the tension at each control point, effectively controlling the curvature of the curve.
4. **Parameterization**: Parameters such as knot vectors and tension values are adjusted to fine-tune the spline curve and improve its fidelity to the original ECG waveform. This allows for greater control over the interpolation process and ensures that the generated curve closely matches the underlying cardiac activity.
5. **Synthetic ECG Signal Generation**: Once the spline curve is optimized, it is used to generate synthetic ECG signals by interpolating between the control points. These synthetic signals can mimic both normal and abnormal ECG patterns, providing a valuable tool for research, algorithm development, and educational purposes [24–26].

Overall, parametric quartic splines offer a flexible and efficient method for ECG modeling, allowing for the creation of realistic synthetic signals that closely resemble real-world data.

1.2.1 Breakdown of the essential requirements for ECG modeling

Synthetic ECG generation serves several valuable purposes in the field of biomedical engineering and cardiology:

Data Augmentation: Synthetic ECG generation allows researchers to augment existing datasets by generating additional samples with varying characteristics. This augmented dataset can enhance the robustness and generalizability of machine learning models trained on limited data.

Anomaly Detection: By generating synthetic ECG signals with known anomalies or abnormalities, researchers can train machine learning models to detect and classify these conditions accurately. This enables the development of automated diagnostic tools for early detection of cardiac abnormalities.

Model Validation: Synthetic ECG signals provide a controlled environment for testing and validating ECG analysis algorithms and models. Researchers can simulate various physiological scenarios and evaluate the performance of their algorithms under different conditions.

Privacy Preservation: Generating synthetic ECG data can help address privacy concerns associated with sharing real patient data. Synthetic datasets can be used for algorithm development and testing without compromising patient privacy or confidentiality.

Education and Training: Synthetic ECG signals are valuable for educational purposes, allowing students and healthcare professionals to practice ECG interpretation skills in a risk-free environment. Synthetic datasets can simulate a wide range of cardiac conditions, providing learners with diverse learning experiences.

Research Exploration: Synthetic ECG generation opens up opportunities for exploring novel research questions and hypotheses in cardiology and biomedical engineering. Researchers can investigate the impact of specific ECG features on diagnostic accuracy or study the effectiveness of different signal processing techniques [14–16].

Overall, synthetic ECG generation plays a crucial role in advancing research and innovation in cardiology, facilitating the development of more accurate diagnostic tools, predictive models, and personalized healthcare solutions.

1.3 RESULT

To apply the spline method for modeling ECG signals, we utilized real ECG data obtained from the PhysioNet database. PhysioNet, managed by the MIT Laboratory for Computational Physiology, offers a repository of freely available medical research data [27]. Thirteen critical points crucial for ECG signal analysis were selected for modeling based on clinical significance [14, 15]. These points were identified according to the unique characteristics of ECG signals, and their locations by star marks are depicted in Figure 1.1. Figure 1.1 visually illustrates the comparison between the original ECG signal and the modeled ECG signal. This comparison vividly demonstrates the close alignment between the modeled signal and the original signal. In the figure, the real ECG signal is denoted by star markers, while the modeled ECG is depicted by a solid line. This graphical representation underscores the fidelity of the modeled signal in faithfully following the original ECG waveform.

The visual comparison of original signal and modelled signal is also shown in Figure 1.1. It indicates that how modelled signal closely follow the original signal. The real ECG is represented by star marker and modelled ECG represented by a solid line. This process was iterated for both normal and abnormal ECG signals, resulting in the creation of a comprehensive ECG signal database tailored for machine learning classification purposes. The database comprised 168 instances. Subsequently, three distinct machine learning techniques were employed: K-nearest neighbors (KNN), Naïve Bayes, and AdaBoost. Each technique was applied to the prepared dataset to assess its classification performance. The classification

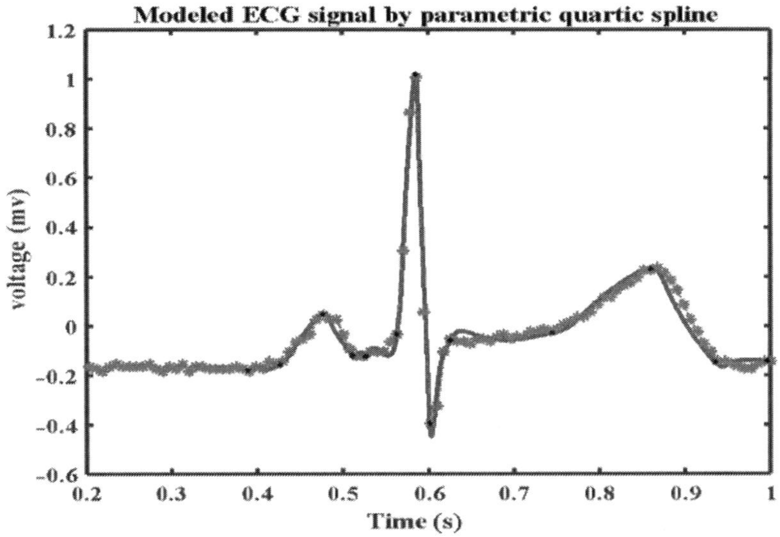

Figure 1.1 Location of critical points and original ECG signal along with the modelled ECG signal.

Table 1.1 Scores of machine learning classification technique

Model	*AUC*	*CA*	*F1*	*Precision*	*Recall*
kNN	0.9887	0.9482	0.9483	0.9484	0.9482
Naive Bayes	0.9999	0.9931	0.9931	0.9932	0.9931
AdaBoost	0.9773	0.9827	0.9826	0.9828	0.9827

accuracies obtained were as follows: 94% for KNN, 99% for Naïve Bayes, and 98% for AdaBoost.

The Orange software platform was utilized to implement these machine learning techniques. Renowned for its user-friendly interface designed for data analysis and machine learning tasks, Orange facilitated seamless execution of the methods. Notably, among the employed techniques, Naïve Bayes demonstrated superior performance. This was evidenced by its higher classification accuracy compared to KNN and AdaBoost. To provide further insight into the performance of each method, confusion matrices were generated. These matrices, presented in Tables 1.1, 1.2, 1.3, and 1.4, offer a detailed breakdown of the classification outcomes, shedding light on the strengths and weaknesses of each approach.

Table 1.2 Confusion matrix for Naïve Bayes

Predicted				
		"A"	"N"	Σ
Actual	"A"	100.0 %	2.0 %	**190**
	"N"	0.0 %	98.0 %	**100**
	Σ	**188**	**102**	**290**

Table 1.3 Confusion matrix for KNN

Predicted				
		"A"	"N"	Σ
Actual	"A"	96.3 %	7.9 %	**190**
	"N"	3.7 %	92.1 %	**100**
	Σ	**189**	**101**	**290**

Table 1.4 Confusion Matrix for AdaBoost

Predicted				
		"A"	"N"	Σ
Actual	"A"	97.9 %	1.0 %	**190**
	"N"	2.1 %	99.0 %	**100**
	Σ	**193**	**97**	**290**

1.4 CONCLUSION

The pivotal role of synthetic ECG generation in propelling advancements in cardiology research and innovation. By harnessing this technique, researchers can propel the development of more precise diagnostic tools, predictive models, and personalized healthcare solutions tailored to individual patient needs. The focus of the presented work has been on ECG modeling utilizing a parametric quartic spline method, complemented by machine learning classification applied to the resultant database. Through this approach, the study aimed to enhance our understanding of ECG signals and improve automated classification capabilities.

Machine learning techniques, including KNN, Naïve Bayes, and AdaBoost, were employed to analyze the derived ECG database. These methods demonstrated considerable potential in automating classification tasks, thereby streamlining diagnostic processes. Notably, among the three techniques, Naïve Bayes emerged as particularly promising, exhibiting superior accuracy compared to its counterparts. Overall, the findings underscore the effectiveness of combining ECG modeling with machine learning classification techniques. This integrated approach holds significant promise for revolutionizing cardiac healthcare by enabling more accurate diagnoses, predictive insights, and tailored treatment strategies. As such, continued exploration and refinement of these methodologies are imperative for driving future advancements in the field of cardiology.

Declarations

Ethical approval: Not applicable.

Competing interests: The authors have no competing interests

Funding: Not applicable.

Availability of data and material: The datasets can be obtained from the corresponding author on an appropriate request.

Code availability: The code can be obtained from the corresponding author on an appropriate request.

REFERENCES

[1] Mishra, A., & Bhusnur, S. (2022). Recapitulation of synthetic ECG signal generation methods and analysis. *International Journal of Signal Processing Systems*, *10*, 14–17.

[2] Agrawal, A. K., Mishra, A., & Kumar, R. (2014). ECG signal compression using efficient transformations. *International Journal of Innovations in Engineering and Technology (IJIET)*, *4*(1), 13–19.

[3] Dössel, O., Luongo, G., Nagel, C., & Loewe, A. (2021). Computer modeling of the heart for ECG interpretation—a review. *Hearts*, *2*(3), 350–368.

[4] Khaleelu Rehman, B., Kumar, A., & Sharma, P. (2018). Modeling and simulation of ECG signal for heartbeat application. In *Intelligent Communication, Control and Devices: Proceedings of ICICCD 2017* (pp. 503–511). Springer Singapore.

[5] Mishra, A., & Bhusnur, S. (2022, October). A new adaptive modeling and denoising of real ECG signal. In *2022 IEEE 3rd Global Conference for Advancement in Technology (GCAT)* (pp. 1–6). IEEE.

[6] Wilber, H., Damle, A., & Townsend, A. (2022). Data-driven algorithms for signal processing with trigonometric rational functions. *SIAM Journal on Scientific Computing*, *44*(3), C185–C209.

[7] Kundu, P. G. (2015). Electrocardiogram synthesis using Gaussian and Fourier models. In *IEEE International Conference on Research in Computational Intelligence and Communication Networks (ICRCICN)*, (pp. 312–317).

[8] Delaney, A. M., Brophy, E., & Ward, T. E. (2019). Synthesis of realistic ECG using generative adversarial networks. arXiv preprint arXiv:1909.09150.

[9] Hazra, D., & Byun, Y. C. (2020). SynSigGAN: generative adversarial networks for synthetic biomedical signal generation. *Biology*, *9*(12), 441.

[10] Wulan, N., Wang, W., Sun, P., Wang, K., Xia, Y., & Zhang, H. (2020). Generating electrocardiogram signals by deep learning. *Neurocomputing*, *404*, 122–136.

[11] Edmond Adib, F. A., & Prevost, J. J. (2021). Synthetic ECG signal generation using generative neural networks. *PloS One*, *20*(3), e0271270.

[12] Gerasimov, A. K., & Pedonova, Z. N. (2022). Development of hardware and software for generating test ECG signals. *Biomedical Engineering*, *55*(5), 315–319.

[13] Dolinský, P., Andráš, I., Michaeli, L., & Šaliga, J., 2020. An ECG signal model based on a parametric description of the characteristic waves. *Acta IMEKO*, *9*(2). https://acta.imeko.org/index.php/acta-imeko/article/view/IMEKO-ACTA-09%20%282020%29-02-02

[14] Mishra, A., Bhusnur, S., & Mishra, S. (2023). A neoteric parametric representation and classification of ECG signal. *Circuits, Systems, and Signal Processing*, *42*(9), 5725–5738.

[15] Mishra, A., & Bhusnur, S. (2023). A piecewise spline approach for modeling of ECG signals. *Biomedical Physics & Engineering Express*, *9*(6), 065017.

[16] Mishra, A., Bhusnur, S., & Mishra, S. K. (2022). Advancing health sciences and biomedical technology: A parametric spline based ECG modeling. *Everyman's Science*, *57*(2).

[17] Mishra, A., Bhusnur, S., Mishra, S. K., & Singh, P. (2023). Innovative ECG signal modeling: Unleashing the potential of the parametric spline approach for advancing global health challenges. *Journal of Technical Education*, *46*, 160.

[18] Mathews, S. M., Kambhamettu, C., and Barner, K. E. (2018). "A novel application of deep learning for single-lead ECG classification." *Computers in Biology and Medicine*, *99*, 53–62.

[19] Ebrahimi, Z., Loni, M., Daneshtalab, M., & Gharehbaghi, A. (2020). A review on deep learning methods for ECG arrhythmia classification. *Expert Systems with Applications: X*, *7*, 100033.

[20] Houssein, E. H., Kilany, M., & Hassanien, A. E. (2017). ECG signals classification: a review. *International Journal of Intelligent Engineering Informatics*, *5*(4), 376–396.

[21] Jambukia, S. H., Dabhi, V. K., & Prajapati, H. B. (2015). Classification of ECG signals using machine learning techniques: a survey. In *2015*

International Conference on Advances in Computer Engineering and Applications (pp. 714–721). IEEE.

[22] Ishak, A., Ginting, A., Siregar, K., & Junika, C. (2020, December). Classification of heart disease using decision tree algorithm. In *IOP Conference Series: Materials Science and Engineering* (Vol. 1003, No. 1, p. 012119). IOP Publishing.

[23] Selek, M. B., Yesilkaya, B., Egeli, S. S., & Isler, Y. (2021). The effect of principal component analysis in the diagnosis of congestive heart failure via heart rate variability analysis. *Proceedings of the Institution of Mechanical Engineers, Part H: Journal of Engineering in Medicine*, *235*(12), 1479–1488.

[24] Ibrahim, Z. (1991). *CAD/CAM Theory and Practice*. McGraw-Hill Higher Education.

[25] Li, J., & Liu, C. (2022). Cubic trigonometric Hermite interpolation curve: construction, properties, and shape optimization. *Journal of Function Spaces*, *2022*(1), 7525056

[26] Mishra, A., Bhusnur, S., Mishra, S. K., & Singh, P. (2024). Exploring a new frontier in cardiac diagnosis: ECG analysis enhanced by machine learning and parametric quartic spline modeling. *Journal of Electrocardiology*, *85*, 19–24.

[27] Goldberger, A. L., Amaral, L. A., Glass, L., Hausdorff, J. M., Ivanov, P. C., Mark, R. G., Mietus, J. E., Moody, G. B., Peng, C. K., & Stanley, H. E. (2000). PhysioBank, PhysioToolkit, and PhysioNet: components of a new research resource for complex physiologic signals. *Circulation*, *101*(23), 215–e220.

Chapter 2

Quantum networking paradigm

Preety Aneja

2.1 INTRODUCTION

Our universe shows a fascinating behaviour at the microscopic level, and there are many occurrences in nature which are hard to be explained. Have we tried to think why a quantum computer can perform better as compared to a typical computer? And what are the unique features it possesses? All these questions have their answers hidden in understanding the fundamentals of nature. The branch of physics, known as quantum mechanics (Konishi and Paffuti 2009), explains the behaviour and dynamics of particles at the microscopic level.

As the research and advancements in quantum mechanics grow rapidly, the existence of quantum computers, quantum networks and now, the quantum internet has become possible. There are significant advantages of quantum computing over traditional or classical computing. Quantum computing harness the fundamental concepts of quantum mechanics like entanglement, superposition, etc (Wineland 2013), which offer significant advantages over traditional computing methods to solve the complex problems like logistics, drug designing, banking, statistical science, secure communication, etc. The realization of quantum computers has become possible only with the remarkable developments in designing quantum software, quantum hardware research, quantum algorithm creation, etc.

The fundamental unit of information in classical computers is a binary digit. Each binary digit can exist in two states, that is either 0 or 1. On the other hand, the basic unit of quantum information in case of quantum computers is a qubit. The qubits can be prepared in various physical systems such as electronic energy states of atoms, optical degrees of freedom of photons, spin states of electrons or nuclei, etc. (Bennett and DiVincenzo 2000; Bouwmeester and Zeilinger 2000). The properties of qubits are generally based on physical properties of the particle, such as the spin of an electron, and hence it can differ in various ways. It is known that an electron can have either up spin or down spin, but in quantum mechanics, there is a probability for an electron to exist in a state which has a spin in both

 DOI: 10.1201/9781003569619-2

directions simultaneously. This concept is known as the superposition of two states in quantum mechanics. This concept sounds strange to our general observation in accordance with classical physics. However, this type of peculiar behaviour shown by particles would be helpful in various ways. For instance, qubits can store much more information than conventional bits. A classical bit can take values as 0 and 1 only. With two bits, there exist four possible combinations: 00, 01, 10 or 11. While in quantum computing, a qubit can take value 0 and 1 simultaneously due to superposition property. Thus, with two qubits, all four combinations (00, 01, 10, 11) can exist at once. This property in quantum computing has increased the amount of information stored in qubits tremendously.

The operations performed by quantum computers are even faster than the fastest supercomputers. Quantum computing provides enhanced security in communication systems also. But then the question arises, what makes the transmission of quantum information possible? Quantum information, in the form of qubits, travels from one quantum processor to other processor over the long distances. This is possible through quantum networks which form an essential part of quantum communication. Quantum computers can provide enhanced security and could be potential threats to current encryption methods (Klima et al. 2018). As the information travels through qubits, any attempt of tampering with the information can affect the quantum states of the particles involved. Thus, quantum networks offer unprecedented capabilities in secure communication.

Further, quantum memories play a crucial role in quantum networks and hence quantum communication which preserves the unique properties of qubits. These technologies are advancing and evolving rapidly in revolutionizing the realm of computation, data processing and transmission and cybersecurity. The goal of achieving quantum supremacy is becoming reality in today's era since various companies and research institutions have already designed and developed quantum computers with over 50 qubits. The computation time taken by these quantum computers to solve complex problems has been reduced in seconds which otherwise would be in years if solved by traditional computers. The future of quantum computing and hence quantum networking is very bright.

This chapter provides a fresh perspective on the quantum networks by examining the various quantum elements like quantum computers, qubits, quantum circuits algorithms, etc. as an interconnected system rather than isolated components so as to make a cohesive quantum network. This chapter highlights the integration of various quantum technologies to transform existing communication frameworks and provides a roadmap to achieve quantum internet with enhanced security over classical internet. Section 2.2 covers the detailed discussion on fundamentals of quantum networks, focussing on quantum computing and the unique properties of superposition, entanglement, interference etc. Section 2.3 discusses about the quantum network paradigms like QKD, quantum teleportation, etc. and their pivotal role in future communication systems. Section 2.4 covers the applications and current state of art studies of quantum networking. Sections 2.5 and 2.6 will discuss the opportunities and challenges of integrating quantum computing in quantum-based network paradigms. The last section concludes the chapter and explores the future directions.

2.2 FUNDAMENTALS OF QUANTUM NETWORKS

Quantum networks are systems designed to transmit and process information based on the principles of quantum mechanics rather than classical physics. This section explores the essential components and principles behind quantum networks, setting the groundwork for understanding how these systems enable secure communication, distributed computing, and advanced quantum sensing as follows:

2.2.1 Quantum computing and qubits

The principles of quantum mechanics form the basis of quantum computing to perform computation in a way that is different from classical computing at the fundamental level. The operation of quantum computers is dependent on the quantum effects like superposition and entanglement in a similar way as the operation of classical computers relies on the behaviour of electrons in semiconductor materials (Langlois 2002). In traditional digital computing, the fundamental unit of information is a binary digit which can take two values, 0 or 1. In quantum computing, the fundamental unit is called qubit (Figure 2.1) which may take value '0' or '1' or can exist in the superposition of these two states simultaneously (Nielsen and Chuang 2010; Nadj-Perge et al. 2010; Hendrickx et al. 2020). Quantum computing has drastically revolutionized the field of computation technology by exploiting the unique principles of quantum mechanics so as to surpass the specific tasks performed by classical computers. Quantum computing has many applications in various domains such as cryptography, materials science, healthcare, logistics, climate modelling, etc.

2.2.2 Superposition

In quantum mechanics, the superposition is a unique property which is exhibited by the qubits. This means a qubit can exist simultaneously in the superposition of states '0' or '1' unlike the classical bits which can only take the values 0 and 1 at once (Figure 2.2). Mathematically, this property can be represented as

$$|\psi\rangle = a|0\rangle + b|1\rangle,$$

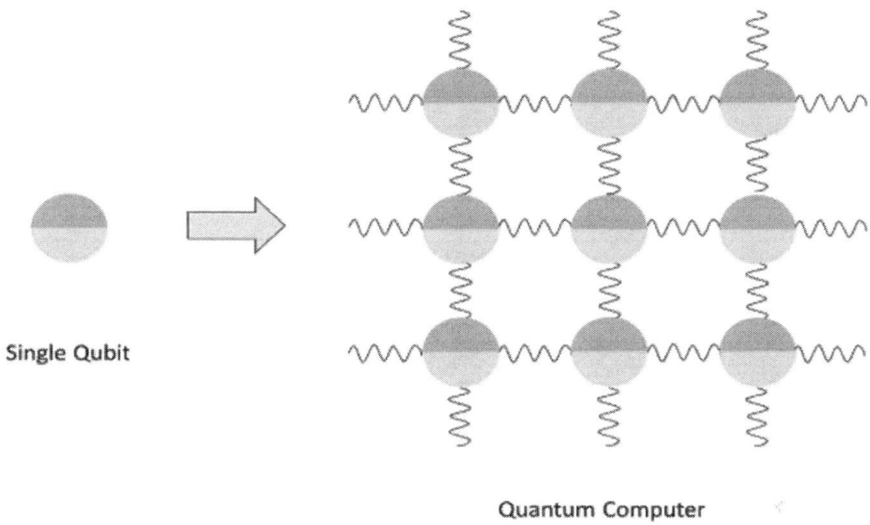

Figure 2.1 From qubits to quantum computer.

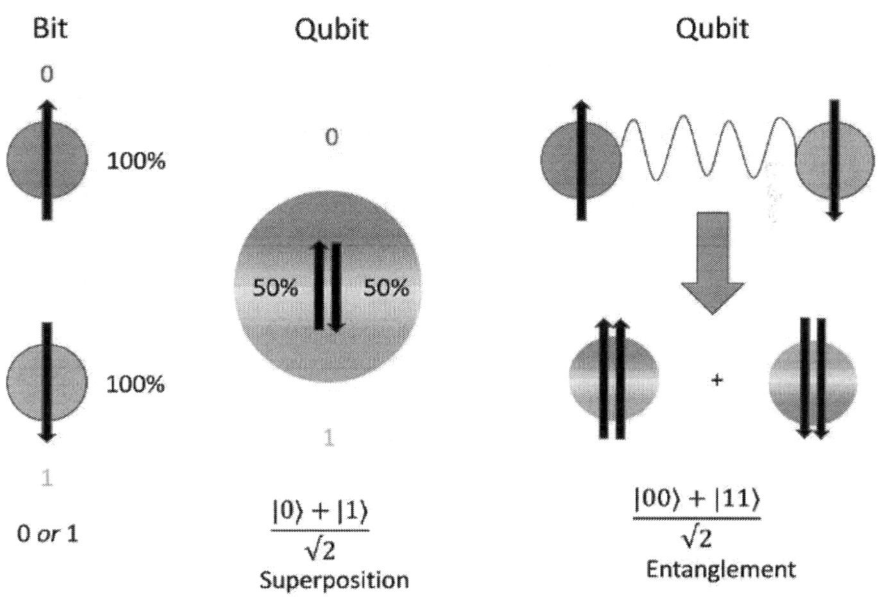

Figure 2.2 Illustration of a bit and qubit.

where $|\psi\rangle$ is a wave function, a and b are complex numbers representing the probability amplitudes of the respective states (Nielsen and Chuang 2010). The magnitude squared of these probability amplitudes will give us the probability of the qubit being either in a particular state and must satisfy the no

$$|a|^2 + |b|^2 = 1.$$

Thus, quantum computers can store and process a vast amount of information due to this unique property of superposition as a huge computational space becomes available. For a system with N number of qubits, the computational domain increases exponentially as 2^N. The exponential growth shows that even with a small number of qubits, a large amount of information can be stored and processed by the quantum computers. This feature of quantum computers has enabled them to solve many complex problems like factoring large numbers, simulation of quantum systems, etc., in contrast to classical computing methods.

2.2.3 Entanglement

Entanglement is another unique property exploited in quantum computing (Bennett and DiVincenzo 2000). Qubits can exist in an entangled state unlike the classical bits, which are totally independent of each other. This entangled state of qubits creates a direct link between two qubits globally even if the two qubits are physically apart. This means if one of the two qubits is probed, the other will get affected due to entanglement. The measurement on one entangled qubit will instantaneously influence the state of other qubit even if these qubits are far apart from each other. Thus, the qubits in the entangled state will have a correlation which classical bits do not exhibit. The entangled system of the qubits must be described as a whole rather than independently (Figure 2.2). Mathematically, the entangled state of the two qubits is represented as

$$|\psi\rangle = a|00\rangle + b|01\rangle + c|10\rangle + d|11\rangle,$$

where a, b, c and d are complex probability amplitudes. This interconnection of multiple qubits over long distances become a valuable asset in quantum computing which has applications in dense coding, quantum teleportation, quantum key distribution (QKD) and quantum simulations (Gill 2021). The correlation among entangled qubits cannot be explained by classical physics and, thus, can lead to more complex and faster computations.

2.2.4 Quantum interference

Quantum states can possibly exhibit the interference phenomena in a similar way as the waves interfere, which leads to maxima and minima (Kipnis 2012). Quantum interference can lead to the amplification of certain outcomes while cancelling the other analogous to constructive and destructive interference in wave theory. The principle of quantum interference is useful in quantum algorithms to enhance the probability of correct solutions as compared to the probability of incorrect solutions, which get diminished. This property can enhance the computational power and efficiency over classical technology.

2.2.5 Measurement

The last stage in quantum computation is measurement, thus ultimately collapsing the qubit from a superposition state into one of the definite states, 0 or 1. The process of measurement ultimately leads to a classical output from a quantum system by destroying the superposition state of qubits to a definite state. Correct outcome is typically guaranteed by the quantum algorithms but stochastic nature of the process can even lead to the non-sampling of the correct outcome. Thus, some classical technique is required after data processing to obtain the final correct output from raw results with quantum computer.

2.2.6 Quantum gates and circuits

The quantum versions of classical gates used to carry out the quantum computation are called quantum gates. Quantum gates operate on superpositions of states unlike classical gates and thus enabling quantum parallelism. Quantum gates operate on qubits through unitary operations to manipulate the state of qubits while preserving quantum properties like superposition, entanglement etc. Quantum gates are reversible and are able to perform reverse calculations and complex computations. These gates can be single-qubit as well as multi-qubits. Examples include single-qubit gates such as Pauli-X (analogous to the classical NOT gate), Pauli-Y, Pauli-Z, Hadamard (H) and Phase (S) gates; two-qubit gates, such as the Controlled-NOT (CNOT) gate, which flips the state of a qubit, thus enabling operations that are impossible with classical gates alone. A sequence of quantum gates can perform complex algorithms and thus forms a quantum circuit.

2.2.7 Quantum algorithms

Quantum circuits form the foundation for implementing quantum algorithms, which are then applied to one or more qubits to solve specific problems. Some of the examples of quantum algorithms are Shor's algorithm for factorization of integers, Grover's algorithm for searching databases, etc. Both these algorithms provide faster speedups in comparison to classical algorithms (Shor 1999; Grover 1996).

2.3 QUANTUM-BASED NETWORKS PARADIGMS: THE NEXT FRONTIER IN QUANTUM COMMUNICATION

Although universal quantum computers or simulators are still in development, efforts are being made by researchers in linking these quantum computers to build quantum networks. Quantum based network paradigms are an emerging field at the intersection of quantum physics and information technology to bring a transformative impact in computation, communication and data transmission with unparalleled security and computational power. Each paradigm has a unique role, mechanism, applications, challenges and advantages. The successful implementation of these paradigms will play a crucial role in the future of communication networks and information processing at the global level. Quantum networks can be better understood by discussing its elements as follows:

- Quantum networks are facilitated with quantum processors with minimum one qubit. A small quantum computer is known as a quantum processor which can perform quantum logic gates on qubits. The application is run on these quantum processors with one qubit known as end nodes. It might be possible to have quantum processors with more than one qubit as per the requirement of the application.
- Communication lines are required for the transfer of qubits and telecom wires can be utilized for this purpose.
- In quantum networks, qubits are delivered to the required processors with the help of optical switches in order to optimizing the communication process.
- Fourth, long distance transmission of qubits is possible through quantum repeaters so as to avoid losses (Gu et al. 2024).

Key aspects and technologies in this domain include:

2.3.1 Quantum key distribution

QKD network is a typical example of a quantum network which does not require entangling the quantum nodes. Unlike classical cryptographic methods, QKD is based on information theoretical security utilizing fundamental laws of quantum mechanics and thus making it secure against any future advancement in computational power (Xu et al. 2020; Zhang et al.

2018). In a QKD network, quantum channels (optical fibres or free space links) are used for the distribution of quantum keys between two parties. These keys are used to encrypt and decrypt messages transmitted through the communication channel. Any attempt at eavesdropping is detectable due to the disturbance of quantum states in the act of quantum measurement. Bennett and Brassard (1984) proposed BB84 protocol which uses four polarization states of photons for encoding bits. Later on, more and more protocols were proposed and developed like E91, etc (Zhang et al. 2024).

2.3.2 Quantum teleportation

It is a process in which the quantum state of a particle is allowed to be transmitted directly from one location to another, without physically transporting the particle itself with the help of classical communication (Olmschenk et al. 2009). This quantum network uses entangled quantum nodes in which the qubits can be processed and transmitted directly. This paradigm creates a robust and versatile network offering incomparable security and computational power (Devulapalli et al. 2024).

2.3.3 Cloud-based quantum computing

Cloud-based quantum computing provides access to quantum emulators and processors over the internet (Soeparno and Perbangsa 2021). As a result, users can perform quantum computations instead of using local quantum hardware. This paradigm of quantum networking has broadened the access of quantum computing resources and thus speed up more research and advancements in this field (Golec et al. 2024).

2.3.4 Entanglement-enabled communication

Entanglement-enabled communication makes use of qubits in entangled state to establish communication links with enhanced security. Any attempt at interference or eavesdropping is detectable as ensured by this network paradigm which in turn provides a robust foundation for secure data transmission (Nötzel 2020).

2.3.5 Quantum repeaters

Quantum repeaters can extend the range of quantum communication by correcting for losses and errors due to decoherence or noise without directly measuring the quantum information (Azuma et al. 2023). These are used for enabling the reliable transmission of quantum information over longer distances without significant loss or degradation of signal. In local area networks or even in some metro area networks with short distances, quantum repeaters are not required as the qubit fidelity can be maintained easily. However, quantum repeaters become essential to implement as the quantum networks scale up to cover some larger geographic area over a country, region such as wide area networks (WANs), some larger metro networks etc. The working of quantum repeaters involves the breaking of the transmission path into shorter segments so that qubit fidelity can be maintained over these segments. Pairs of entangled qubits are created across each segment. Entanglement between adjacent segments is created and then entanglement is swapped over the longer distances to maintain the purity and integrity of quantum states. Quantum memory is a core component of quantum repeaters as it can store and release the qubit at a later time. This will enable the synchronization of entanglement over different segments (Azuma et al. 2023).

2.3.6 Quantum internet

The internet is playing an integral role in connecting people across globe and hence, become an important part in our daily lives. However, the current internet infrastructure is more prone to cyber-attacks posing a threat to sensitive information (Aslan et al. 2023; Carlo and Obergfaell 2024). Quantum internet can solve this problem. The quantum internet is a theoretical concept of interconnected quantum computers and other quantum devices such that information can be sent and received using quantum signals (Wehner et al. 2018). Many complex problems could be solved with the help of quantum internet in co-existence with the classical internet. The motivation behind quantum internet comes from the transformative impact of classical internet on our society. The development of quantum internet can have profound implications for security, scientific discovery and computing. The quantum internet can span various domains, including metrology, distributed quantum computing and quantum-enhanced secure communication potentially revolutionizing areas such as financial risk analysis, data encryption and material science (Cacciapuoti et al. 2019).

2.4 PRACTICAL APPLICATIONS AND STATE-OF-ART STUDIES IN QUANTUM NETWORKING

Till date, no one has managed to develop a large-scale and sustained quantum network, but there are many advancements in this direction as discussed in Table 2.1.

2.5 OPPORTUNITIES OF QUANTUM COMPUTING IN QUANTUM-BASED NETWORK PARADIGMS

Integrating quantum computing into quantum-based network paradigms is a groundbreaking development that has the potential to revolutionize information processing. Quantum computing can significantly enhance the

Table 2.1 Applications of quantum networking

Year	*Breakthroughs in quantum networking*	*Organizations involved*	*Description*
2016	Micius Satellite	Chinese Academy of Sciences	Launched by China, it achieved quantum key distribution (QKD) over 4,600 km (Jianlan 2016)
2017	Satellite-Based Quantum Network	University of Science and Technology of China	Scientists demonstrated by transmitting the entangled photons between a satellite and ground stations over 700 miles. The main drawback of this experiment was low recovery rate of photons i.e. only one photon out of every 6 million was recovered (Nellis n.d., Yin et al. 2017)
2017	Beijing-Shanghai Quantum Communication Network	China Telecom, National University of Defense Technology	World's first long-distance quantum-secured communication fibre optic network over 2,000 km for integrating QKD. Enhances data security for sensitive information in finance and government sectors (Na 2017)
2019	Brookhaven National Laboratory Quantum Network	Brookhaven National Laboratory, Stony Brook University, U.S. Department of Energy	Used portable quantum entanglement sources and fibre optics to achieve entanglement over 10 miles, later expanded to 80 miles (Nellis n.d.)
2019	Quantum Internet Alliance	Multiple European universities and research institutions	A collaborative project initiated to develop a quantum internet in Europe. Promotes research into quantum communication nodes and repeaters for building a functional quantum internet (Quantum Internet Alliance n.d.)
2020	52-Mile Quantum Loop	University of Chicago, Argonne National Laboratory	Demonstrated the core functionality of a quantum network to achieve success in testing a 52-mile quantum loop by using existing fibre optic cables (Lerner 2020)
2021	Caltech and NASA Quantum Communication Testbed	Caltech, NASA	Developed a testbed for secure information transmission in fibre optics and free-space optical links. Explores practical applications of quantum networking for secure communications in space exploration (CQNET 2020)

(*continued*)

Table 2.1 (Cont.)

Year	*Breakthroughs in quantum networking*	*Organizations involved*	*Description*
2022	Expanded Chicago Quantum Network	University of Chicago, Argonne National Laboratory	Expanded by 35 miles, making it one of the longest networks in the nation with approximately 124 miles of optical fibre. Enhances the capability and reach of quantum networking in the United States (Hu 2022)
2022	Swiss Quantum Initiative	Swiss Federal Institute of Technology (ETH Zurich)	Implementation of the first national quantum communication network in Switzerland, connecting key institutions. Facilitates secure data sharing and research collaboration across institutions, enhancing national cybersecurity (Swiss Quantum Initiative n.d.)
2022	Quantum-Secured 5G Networks	University of Bristol, multiple telecom companies	Enhances security protocols for 5G networks, protecting against potential cyber threats using quantum techniques (Quantum Secured 5G Network n.d.)
2022	Jinan1 satellite	Chinese Academy of Sciences	It was second quantum satellite which can generate quantum keys at speeds two or three times faster than Micius satellite but weight is only one-sixth of Micius's weight (Chen 2022)

capabilities of quantum networks, leading to more secure and efficient communication. Let's discuss some of the opportunities:

1. *Enhanced computational power*: Quantum computing can perform complex calculations at faster speeds than classical computers (Farhi et al. 2014). This characteristic feature of quantum computers can bring transformative impact on industries such as cryptography, optimization and drug discovery.
2. *Enhanced Security*: Quantum networks are based on the fundamental principles of quantum mechanics ensuring that any attempt to intercept information can be readily detected. Quantum networks like QKD had greatly enhanced the security in communication networks (von Burg et al. 2021). Quantum computing has enabled more sophisticated encryption methods which are virtually unbreakable by classical means to improve the security features.

3. *Efficient information processing*: Quantum algorithms can optimize network protocols and have the tendency to reduce delay in the data transmission (Gisin et al. 2002; Djordjevic 2012). This will overall increase the capacity of quantum networks to transmit data with minimum delays. Quantum computing and hence quantum networks can lead to more efficient processing and transmission of data.
4. *Quantum Sensors*: Quantum networks aim to interconnect the quantum devices over a long distance and hence find applications from medical imaging to environmental monitoring with enhanced sensing capabilities (Landman et al. 2022).
5. *New paradigms in network design*: The advancements in quantum physics had led to the development of novel network architectures based on it. For instance, communication systems are greatly affected by the development of quantum repeaters and quantum routers which can extend the communication range and hence, reliability of quantum communication.
6. *Scientific Research*: Quantum networks can expedite the collaborative research by enabling shared access to quantum resources and linking quantum laboratories. Overall, quantum-based network paradigms hold the potential to revolutionize communication and computation that are unattainable with classical technologies.

2.6 CHALLENGES

The role of quantum computing in quantum-based networks is exciting. There exist challenges also as discussed below:

1. *Quantum decoherence*: The decay of quantum information as a result of interaction with the environment is called quantum decoherence (Schlosshauer 2019). The quantum states are very much prone to decoherence on interaction with environment. This can lead to the loss of information so the main challenge is to protect the purity of quantum states by developing advanced isolation techniques and materials.
2. *Quantum and Classical integration*: The integration of quantum networks with the existing classical networks so as to create hybrid systems brings new challenges (Shi and Malaney 2023; Reilly 2019). The integration includes designing interfaces and protocols for the seamless interaction between quantum and classical systems (DiAdamo et al. 2022).
3. *Atmospheric Interference*: Free-space quantum communication is greatly affected by weather conditions and atmospheric turbulence.

4. *Security*: Quantum networks need to be protected against the attacks by continuously monitoring the security protocols, developing safe cryptographic methods and doing rigorous testing.
5. *Development of efficient quantum repeaters*: The quantum repeaters are required to extend the range of quantum communication but robust and efficient repeaters are difficult to build that can perform entanglement swapping and preserve the purity of quantum states without loss of information. Quantum repeaters rely on quantum memories to store and retransmit entangled states. However, current quantum memories struggle with limited coherence times (the duration they can reliably hold quantum states) and low storage efficiency, which restricts network scalability (Dusanowski and Thompson 2023). Further, in large quantum networks, scaling up of numerous repeaters and synchronization between these repeaters is a complex challenge.
6. *Cost and scalability for large-scale integration*: Small quantum networks, while scaling up to larger practical quantum networks require advancements in technology. This includes developing scalable architectures, standardized protocols and modular quantum network components which requires investment and infrastructure also. Quantum hardware is still in development, and real-world problems can be solved only with much more powerful hardware (Monroe and Kim 2013).
7. *Interdisciplinary collaboration*: Research and development in quantum networks requires the expertise in various fields such as quantum physics, computer science, engineering and thus fostering interdisciplinary collaboration.
8. *Standards and protocols*: The development of quantum networks lacks the standardized protocols and regulations for quantum communication.

2.7 CONCLUSION AND FUTURE DIRECTIONS

Quantum computing and hence quantum networks have many real-world applications. With the advent of quantum computing which leverages the principles of quantum mechanics, there occurs a paradigm shift in classical computing since quantum computing enable computations that are infeasible with classical methods. The foundation of quantum computing is based on the qubits and quantum gates, which in turn, offers the unprecedented speed and efficiency in solving complex problems. Thus, quantum-based network paradigms represent a transformative shift in computation, communication technology and information processing to achieve enhanced security, speed and computational power. The potential of quantum computing is exciting as well as challenging also. There exist many challenges with quantum networks such as preservation of entanglement over the long distance, technologies for integrating quantum nodes with existing classical infrastructure, and errors due to decoherence and noise. But the ongoing research and development efforts in this field promise to unlock the full potential of quantum networks to make quantum communication as the cornerstone of global connectivity. Key directions include hybrid quantum-classical protocols to integrate quantum systems with classical infrastructure, standardized communication protocols for interoperability, quantum machine learning (Karthikeyan et al. 2024) and AI-powered network optimization (Khang and Rath 2024) to adapt dynamically to network conditions. It is emphasized that the international collaboration and standards development to realize a global quantum internet infrastructure is urgently needed. A more secure, interconnected, and technologically advanced future can be embarked by embracing quantum-based network paradigms.

REFERENCES

Aslan, Ömer, Semih Serkant Aktuğ, Merve Ozkan-Okay, Abdullah Asim Yilmaz, and Erdal Akin. 2023. "A comprehensive review of cyber security vulnerabilities, threats, attacks, and solutions." *Electronics* 12, no. 6: 1333.

Azuma, Koji, Sophia E. Economou, David Elkouss, Paul Hilaire, Liang Jiang, Hoi-Kwong Lo, and Ilan Tzitrin. 2023. "Quantum repeaters: From quantum networks to the quantum internet." *Reviews of Modern Physics* 95, no. 4: 045006.

Bennett, Charles H., and David P. DiVincenzo. 2000. "Quantum information and computation." *Nature* 404, no. 6775: 247–255.

Bennett, Charles H., and Gilles Brassard. 1984. "Proceedings of the IEEE international conference on computers, systems and signal processing," Bangalore, India, 175–179.

Bouwmeester, Dirk, and Anton Zeilinger. 2000. "The physics of quantum information: basic concepts." In *The physics of quantum information: quantum cryptography, quantum teleportation, quantum computation*, pp. 1–14. Springer Berlin Heidelberg.

Cacciapuoti, Angela Sara, Marcello Caleffi, Francesco Tafuri, Francesco Saverio Cataliotti, Stefano Gherardini, and Giuseppe Bianchi. 2019. "Quantum internet: Networking challenges in distributed quantum computing." *IEEE Network* 34, no. 1: 137–143.

Carlo, Antonio, and Kim Obergfaell. 2024. "Cyberattacks on critical infrastructures and satellite communications." *International Journal of Critical Infrastructure Protection* 46: 100701.

Chen, Stephen. 2022. "China launches new satellite in 'important step' towards global quantum communications network." *SCMP*. Accessed October 28, 2024. www.scmp.com/news/china/science/article/3186802/china-launches-new-satellite-important-step-towards-global.

CQNET. 2020. "Quantum Internet Tested at Caltech and Fermilab." Accessed October 28, 2024. www.caltech.edu/about/news/quantum-internet-tested-caltech-and-fermilab.

Devulapalli, Dhruv, Eddie Schoute, Aniruddha Bapat, Andrew M. Childs, and Alexey V. Gorshkov. 2024. "Quantum routing with teleportation." *Physical Review Research* 6, no. 3: 033313.

DiAdamo, Stephen, Bing Qi, Glen Miller, Ramana Kompella, and Alireza Shabani. 2022. "Packet switching in quantum networks: A path to the quantum internet." *Physical Review Research* 4, no. 4: 043064.

Djordjevic, Ivan. 2012. *Quantum information processing and quantum error correction: an engineering approach*. Academic Press.

Dusanowski, Łukasz, and Jeff D. Thompson. 2023. "A step closer to repeaters for quantum networks." *Nature*. https://doi.org/10.1038/d41586-023-02283-4

Farhi, Edward, Jeffrey Goldstone, and Sam Gutmann. 2014. "A quantum approximate optimization algorithm." *arXiv preprint arXiv:1411.4028*.

Gill, Sukhpal Singh. 2024. "Quantum and blockchain based serverless edge computing: A vision, model, new trends and future directions." *Internet Technology Letters* 7, no. 1: e275.

Gisin, Nicolas, Grégoire Ribordy, Wolfgang Tittel, and Hugo Zbinden. 2002. "Quantum cryptography." *Reviews of Modern Physics* 74, no. 1: 145.

Golec, Muhammed, Emir Sahin Hatay, Mustafa Golec, Murat Uyar, Merve Golec, and Sukhpal Singh Gill. 2024. "Quantum cloud computing: Trends and challenges." *Journal of Economy and Technology* 2: 190–199.

Grover, Lov K. 1996. "A fast quantum mechanical algorithm for database search." In *Proceedings of the twenty-eighth annual ACM symposium on Theory of computing*, pp. 212–219. New York: Association for Computing Machinery.

Gu, Fenglei, Shankar G. Menon, David Maier, Antariksha Das, Tanmoy Chakraborty, Wolfgang Tittel, Hannes Bernien, and Johannes Borregaard. 2024. "Hybrid quantum repeaters with ensemble-based quantum memories and single-spin photon transducers." https://doi.org/10.21203/rs.3.rs-4329745/v1

Hendrickx, N. W., W. I. L. Lawrie, L. Petit, A. Sammak, G. Scappucci, and M. Veldhorst. 2020. "A single-hole spin qubit." *Nature Communications* 11, no. 1: 3478.

Hu, Charlotte. 2022. "Chicago now has a 124-mile quantum network. This is what it's for." Accessed October 28, 2024. www.popsci.com/technology/chicago-quantum-network/.

Jianlan, S. O. N. G. 2016. "Micius Heralds an era of quantum communications." *Bulletin of the Chinese Academy of Sciences* 30, no. 3: 151–154.

Karthikeyan, S., M. Akila, D. Sumathi, and T. Poongodi, eds. 2024. *Quantum machine learning: A modern approach* (1st ed.). Chapman and Hall/CRC. https://doi.org/10.1201/9781003429654

Khang, Alex, and Kali Charan Rath, eds. 2024. *The quantum evolution: Application of AI and robotics in the future of quantum technology*. CRC Press.

Kipnis, Nahum. 2012. *History of the Principle of Interference of Light*. Vol. 5. Birkhäuser.

Klima, Richard E., Richard Klima, Neil P. Sigmon, and Neil Sigmon. 2018. *Cryptology: Classical and modern*. Chapman and Hall/CRC.

Konishi, Kenichi, and Giampiero Paffuti. 2009. *Quantum mechanics: A new introduction*. OUP Oxford.

Landman, Jonas, Natansh Mathur, Yun Yvonna Li, Martin Strahm, Skander Kazdaghli, Anupam Prakash, and Iordanis Kerenidis. 2022. "Quantum methods for neural networks and application to medical image classification." *Quantum* 6: 881.

Langlois, Richard N. 2002. "Computers and semiconductors." In *Technological innovation and economic performance*, pp. 265–284, B. Steil, D.G. Victor and R.R. Nelson (eds.). Princeton, New Jersey: Princeton University Press.

Lerner, Louise. 2020. "Argonne and UChicago scientists take important step in developing national quantum internet." Accessed October 28, 2024. www.anl.gov/article/argonne-and-uchicago-scientists-take-important-step-in-developing-national-quantum-internet.

Monroe, Christopher, and Jungsang Kim. 2013. "Scaling the ion trap quantum processor." *Science* 339, no. 6124: 1164–1169.

Na, Chen. 2017. "Beijing-Shanghai quantum communication network put into use." Accessed October 28, 2024. https://english.cas.cn/newsroom/archive/news_archive/nu2017/201703/t20170324_175288.shtml

Nadj-Perge, S., S. M. Frolov, E. P. A. M. Bakkers, and Leo P. Kouwenhoven. 2010. "Spin–orbit qubit in a semiconductor nanowire." *Nature* 468, no. 7327: 1084–1087.

Nellis, Andrew. n.d. "The quantum internet, explained." Accessed October 28, 2024. https://news.uchicago.edu/explainer/quantum-internet-explained.

Nielsen, Michael A., and Isaac L. Chuang. 2010. *Quantum computation and quantum information*. Cambridge University Press.

Nötzel, Janis. 2020. "Entanglement-enabled communication." *IEEE Journal on Selected Areas in Information Theory* 1, no. 2: 401–415.

Olmschenk, Steven, D. N. Matsukevich, P. Maunz, D. Hayes, L-M. Duan, and C. Monroe. 2009. "Quantum teleportation between distant matter qubits." *Science* 323, no. 5913: 486–489.

Quantum Internet Alliance. n.d. Accessed October 28, 2024. https://qutech.nl/qia/.

Quantum Secured 5G Network. n.d. Accessed October 28, 2024. www.bristol.ac.uk/media-library/sites/engineering/research/smart-internet-lab/Leaflet_Quantum%20Secured_20190120.pdf.

Reilly, D. J. 2019. "Challenges in scaling-up the control interface of a quantum computer." In *2019 IEEE International Electron Devices Meeting (IEDM)*, pp. 31–7. IEEE.

Schlosshauer, Maximilian. 2019. "Quantum decoherence." *Physics Reports* 831: 1–57.

Shi, Wenbo, and Robert Malaney. 2023. "Quantum routing for emerging quantum networks." *IEEE Network 38*(1): 140–146.

Shor, Peter W. 1999. "Polynomial-time algorithms for prime factorization and discrete logarithms on a quantum computer." *SIAM Review* 41, no. 2: 303–332.

Soeparno, Haryono, and Anzaludin Samsinga Perbangsa. 2021. "Cloud quantum computing concept and development: A systematic literature review." *Procedia Computer Science* 179: 944–954.

Swiss Quantum Initiative. n.d. Accessed October 28, 2024. https://qc.ethz.ch/community-and-initiatives/network/swiss-quantum-initiative.html.

von Burg, Vera, Guang Hao Low, Thomas Häner, Damian S. Steiger, Markus Reiher, Martin Roetteler, and Matthias Troyer. 2021. "Quantum computing enhanced computational catalysis." *Physical Review Research* 3, no. 3: 033055.

Wehner, Stephanie, David Elkouss, and Ronald Hanson. 2018. "Quantum internet: A vision for the road ahead." *Science* 362, no. 6412: eaam9288.

Wineland, David J. 2013. "Nobel Lecture: Superposition, entanglement, and raising Schrödinger's cat." *Reviews of Modern Physics* 85, no. 3: 1103–1114.

Xu, Feihu, Xiongfeng Ma, Qiang Zhang, Hoi-Kwong Lo, and Jian-Wei Pan. 2020. "Secure quantum key distribution with realistic devices." *Reviews of Modern Physics* 92, no. 2: 025002.

Yin, Juan, Yuan Cao, Yu-Huai Li, Sheng-Kai Liao, Liang Zhang, Ji-Gang Ren, Wen-Qi Cai et al. 2017. "Satellite-based entanglement distribution over 1200 kilometers." *Science* 356, no. 6343: 1140–1144.

Zhang, Qiang, Feihu Xu, Yu-Ao Chen, Cheng-Zhi Peng, and Jian-Wei Pan. 2018. "Large scale quantum key distribution: challenges and solutions." *Optics Express* 26, no. 18: 24260–24273.

Zhang, Yichen, Yiming Bian, Zhengyu Li, Song Yu, and Hong Guo. 2024. "Continuous-variable quantum key distribution system: Past, present, and future." *Applied Physics Reviews* 11, no. 1: 011318. https://doi.org/10.1063/5.0179566

Chapter 3

Genetic algorithm-based framework for optimizing medical image enhancement

Kirti Aggarwal and Meenal Jain

3.1 INTRODUCTION

Enhancing the clinical capabilities of medical image enhancement is critical for image analysis. The goal of this work is to apply genetic algorithms (GAs) to produce enhanced images. The basic idea entails feeding the initial image through GAs to produce a clear and high-quality reproduced image, which is then enhanced further. This work aims to create a framework for creating the finest possible images by using a cutting-edge algorithm that considers all restrictions to produce better outcomes. For accurate and efficient picture enhancement, it employs the GA. We began with reading an image and preparing the fitness function. This randomly generated image is continuously evolved through many generations, using crossover and mutation, and GA until a superior solution is found. With a dual-peaked intensity distribution, this can enhance the visual impression of a region of interest (ROI) in picture data.

Before processing, image enhancement enhances the quality and information content of the original data. Traditional technique includes contrast enhancement, spatial filtering, and density slicing. In contrast enhancement or stretching, a linear transformation is applied to expand the range of grey levels. Spatial filtering improves linear features that occur naturally, such as fault. Density slicing segments the continuous range of grey tone into a series of density intervals, with each interval represented by a specific color or symbol.

The demand for medical image improvement has expanded in recent years in order to assist clinicians in making accurate diagnoses; hence, medical image processing provides relief to enhance medical images for proper illness diagnosis [1]. One of the primary goals of medical image processing has always been to improve image quality. The clarity of images for human viewing is improved by image enhancement.

Enhancement techniques include removing blur and noise from an image, increasing contrast, and revealing details [2]. While capturing an image, some information is lost, resulting in a blurry or less informative image that

DOI: 10.1201/9781003569619-3

is not always clear. Medical images are poorly lighted, and many key parts of the image are not visible, resulting in visual blur and contrast that is difficult to notice.

Medical research data contains a plethora of information that, in some situations, cannot be simply examined, if at all, using traditional statistical approaches. Physicians who could benefit from using algorithms to solve difficult medical problems are unaware of their existence. As a result, we wanted to use the GA and its medical applications. The employment of the GA in several medical specialties, including radiology, has intriguing implications. This work presents the GA's uses in illness screening, diagnosis, treatment planning, pharmacovigilance, prognosis, and healthcare management, as well as allowing physicians to imagine how they may apply this metaheuristic approach in their own practice. Figure 3.1 represents the flow cycle of GA. The algorithm is defined in detail in Section 3.3.

During this research work, we will conduct extensive research on how GAs work and how they can be used in optimizing the process of image enhancement. We aspire to create a new path for the healthcare sector where the solutions can be achieved using GA. The image enhancement problem can be characterized by several factors, which make the parameter selection problem difficult.

This chapter presents a framework that applies GAs within the context of connective cognition networks to optimize image enhancement across various domains. By harnessing the power of computational intelligence, the framework connects interdisciplinary knowledge—from artificial intelligence to image processing—into an integrated system designed to enhance the quality of images. GAs, as a core component of computational

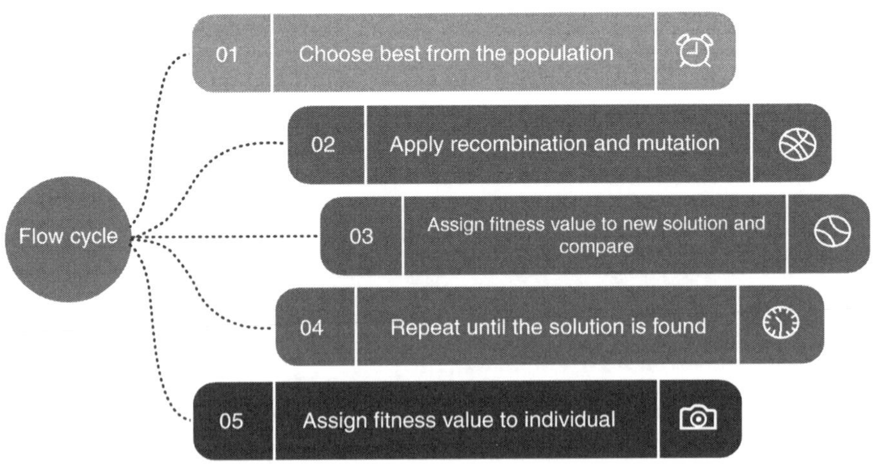

Figure 3.1 Flow cycle of genetic algorithm.

intelligence, are utilized to address the complex optimization challenges involved in image enhancement, providing adaptable and scalable solutions. The chapter highlights how these algorithms mimic evolutionary processes to iteratively improve image quality, reflecting the adaptive nature of connective cognition networks that link diverse cognitive processes and expertise. This approach contributes to advancements in computational intelligence by showcasing how connective cognition can be leveraged to develop innovative solutions for image enhancement, ultimately improving visual outcomes across different applications.

3.2 LITERATURE SURVEY

The majority of extant image enhancement approaches are empirical or heuristic methods that are tied to specific types of images and try to enhance the contrast level of images that have deteriorated throughout the acquisition process. An optimization problem is determining the ideal level of greyscale that enhances each different input image. Some approaches employ Swarm Intelligence techniques. Particle swarm optimization (PSO) is used in multiobjective histogram equalization to improve contrast while preserving brightness. Some works use the utilize similar candidate solution encodings and histogram mapping strategies, all within an optimization framework focused on the Artificial Bee Colony (ABC) algorithm.

In the realm of image processing, nature-inspired optimization strategies are critical. It improves image enhancement, restoration, image segmentation [3], image edge detections, image generation, image fusion, picture pattern recognition, image thresholding, and other functions. So far, numerous optimization strategies for various image processing applications have been proposed. Evolutionary algorithms [4] draw inspiration from biological processes like reproduction, mutation, recombination, and selection. The optimization technique is essential for identifying the most accurate or optimal solutions from a set of possibilities. In scenarios involving multiple individuals, each person may have their own best solution, with the global best representing the top solution among all local bests. Evolutionary algorithms, leveraging fitness functions, have won many challenging problem-solving tasks, and the field that employs these algorithms as problem-solving tools is known as evolutionary computation. The fitness function is a key component of evolutionary computation and enhancing the objective function will result in ideal solutions.

Fireflies are insects that have the inherent ability to illuminate in the dark through flashing and flickering. This biological characteristic inspired the development of the firefly optimization algorithm [5]. Firefly optimization algorithm was used to enhance fuzzy clustering for CT/MR Images [6]. The process of segmentation involves isolating the desired area of interest. Anatomical organs and irregularities such as tumors, cysts, and other anomalies are important in medical imaging for clinicians to diagnose disorders for telemedicine applications. Classic segmentation methods include thresholding, region growth, and edge detection. Clustering, an unsupervised learning approach, groups similar data points together, while fuzzy partitioning integrates comparable pixels based on fuzzy membership value. However, the conventional Fuzzy C-Means (FCM) algorithm is insensitive when it comes to cluster centroid initialization, leading to frequent issues with local minima. Therefore, the optimization process becomes more important during the initialization of cluster centroids, boosting the FCM algorithm's performance. For CT/MR medical picture segmentation, firefly optimization is used with the FCM method in this study.

The nonlinear tensor diffusion filter is used to remove artefacts and denoise the image during the preprocessing stage, effectively reducing computation time by processing fewer pixels overall. When combined with FCM, the firefly optimization produces satisfactory outperforming FCM when paired with other methods such as Cuckoo Search [7], ABC [8], and simulated annealing methods [9]. Cluster validity performance criteria are utilized to determine the optimal number of clusters. In the storage and transmission of medical data, image compression is crucial. For medical photos, lossless compression methods are often used, although lossy vector quantization (VQ) algorithms are employed in scenarios where the reconstructed image quality remains high with an optimal compression ratio. The bat optimization algorithm [10] is inspired by bats' habit of using echolocation to locate prey and navigate through obstacles. In the VQ algorithm, the bat optimization process is used to find the best codebook design. For abdominal CT images, the BAT-VQ compression technique was compared against classical VQ, contextual vector quantization (CVQ), and JPEG lossless systems.

Beni and Wang [11] invented the swarm intelligence algorithm, It is made up of agents or individuals who have a local interest in one another and the environment. They obey individual rules; there is no centralized control system, thus they act independently. Local agent interaction will result in global behavior, which implies global intelligence. Mirjalili et al. [12] introduced this algorithm, based on the leadership and hunting styles of grey wolves. Grey wolves are called apex predators because they are at the top of the food chain. The group members adhere to a highly tight social dominance hierarchy. The first level is alpha, and they are the leaders who make judgments about hunting, sleeping, and other matters. The beta level is the second. These individuals serve as subordinates to the alpha decision-making level. When the alphas become too old or die, the beta candidates will take their position. The third level is known as delta, and it dominates omega. The lowest level of the hierarchy is Omega. Both alpha and beta wolves will use delta wolves as subordinates. The fittest solution in GWO will be alpha, followed by beta, and finally delta. Everything else will be considered Omega. As a result, alpha, beta, delta, and omega will be used to direct hunting.

The ABC optimization method is based on honey bee behavior and was proposed by Karaboga [13]. Three groups of bees are present in this mode. Employed bees, onlooker bees, and scout bees are the three types. Employed bees go on a food hunt and report back to the hive. From the information provided by the employed bees, the observer bees will select the best food source. i.e., each employed bee will have information about a certain food source. Scout bees are employed bees whose food source has been rejected. Kennedy and Eberhart [14] proposed PSO, developed by observing the behavior of fish and birds while they move in groups. The PSO will iterate using the specified parameters until it finds a better solution for the candidate. By taking into account inertia, personal influence, and social impact, a particle, also known as a candidate, can enhance its position. PSO is most commonly used in control systems for functional optimization and optimum control. PSO is used to solve nonlinear optimization issues since it is simple to implement and has fewer parameters. It is adaptable to various uses.

The key findings from our research are that the individuals cannot be created or deleted in PSO. Individuals just move around a terrain that tracks their fitness over time. This is similar to a flock of birds or other animals communicating. Second, fitness proportionate selection and genetic recombination occur in GAs. The PSO equivalents of this are best selection and particle movement. Both methods must be fine-tuned to function in certain problem domains. To achieve faster convergence PSO techniques require normalization of the input vectors. GAs can handle both continuous and discrete values. In terms of computational resources, the GA is quick, simple to deploy, and cost-effective. The ACO is greedier, but it produces superior results, especially when dealing with huge problems. Each method has its own set of mechanics and representational elements. It's crossover and schemata in GAs. Edge-based structure and edge weights are used in ACO. Table 3.1 presents the literature survey summarization.

3.3 GENETIC ALGORITHM

GAs are fundamentally inspired by Charles Darwin's natural selection process, where they accept input and generate output by selecting from various potential solutions. GAs were initially employed by John Holland [15]. The GAs are intended to mimic natural system processes that are required for evolution. To find the best answers, GA searches efficiently in global spaces. GA is more successful at enhancing contrast and generates natural-looking photos. A GA is a search technique inspired by the principles of natural selection mechanism and genetic system. These algorithms are highly parallel and flexible, merging the concept of "survival of the fittest" in string structures with a systematic yet random exchange of information. This

Table 3.1 Literature survey

Study	*Method*	*Application*	*Key findings*
Draa and Bouaziz [2]	Artificial bee colony (ABC) algorithm	Image contrast enhancement	Demonstrated significant improvement in contrast for medical images using a swarm intelligence approach
Qin et al. [3]	Histogram analysis with cloud model	Image segmentation	Used histogram-based techniques to effectively segment medical images, especially for diagnostic applications
Kumar et al. [6]	Firefly optimization + fuzzy clustering	CT/MR image segmentation	Enhanced image segmentation by combining firefly optimization with fuzzy clustering to improve clarity
Rajabioun [7]	Cuckoo optimization algorithm	Medical image processing	Applied cuckoo optimization for efficient processing of high-dimensional medical images
Karaboga and Basturk [8]	Artificial bee colony (ABC)	Functional optimization, control systems	Highlighted ABC's effectiveness in solving nonlinear optimization problems in control systems
Kennedy and Eberhart [14]	Particle swarm optimization (PSO)	Image optimization	Demonstrated PSO's ability to optimize image processing tasks with fewer parameters than other techniques
Karaboga [13]	Artificial bee colony (ABC)	Image optimization	Proposed ABC for handling high-dimensional image data with promising results in image contrast enhancement

approach results in a search strategy that mimics some of the creativity found in human problem-solving. They effectively use previous data to predict and explore new search areas with projected improved outcomes, iteratively applying genetically inspired operators to probable solutions.

Algorithm 1: Pseudo of the Genetic Algorithm

$g = 0$ /* *Initialize Generation counter* */

$P = Initialize\ popultaion$ /* *Random generation of p individuals* */

Repeat UNTIL end condition

$F = Reproducstion(P)$ /* *Selection of p parents thriugh a roulette* */

$S = Crossover(F)$ /* *Swap of information between different pairs of parents* */

$M = Mutation(S) / {}^{*}Alteration\ of\ the\ individual\ bits\ with\ probability\ mc^{*} /$

$P = Updtae\ population(M) / {}^{*}M\ is\ copied\ to\ P(P = M)^{*} /$

$g = g + 1 / {}^{*}Increment\ of\ the\ Generation\ counter\ ^{*} / \pm$

Unlike traditional search strategies, GAs deal with numerous points at once rather than a single point. GAs have been shown to give reliable search in complex spaces both theoretically and experimentally. Pattern recognition and image processing present numerous challenges that require executing efficient searches within complex spaces to identify the optimal solution. GAs use three fundamental procedures on a restricted number of strings to discover the global near-optimal answer. Reproduction/selection, cross-over, and mutation are the three operators.

3.3.1 Parent selection strategies

3.3.1.1 Tournament selection

One of the selection strategies used in GAs is tournament Selection. In this, the top candidates are selected from the present generation. After this, the selected best candidates are advanced to the following generation. In a K-way tournament selection, top k-individuals are selected and they compete against each other [16]. Among the selected candidates only the best candidate is selected and passed on to the next generation. Numerous tournaments are conducted, allowing us to finalize the pick of candidates, those will advance to the following generation.

This selection strategy has a feature called selection pressure, that provides a probabilistic evaluation of a candidate's chance of being chosen for a tournament. If the tournament size is large, weaker candidates will have less likelihood of selection, as they must compete against stronger opponents. The pace of convergence of the GA is determined by the selection pressure parameter. The higher the selection pressure, the higher the rate of convergence. Over a wide range of selection pressures, GAs can uncover best or near-best solutions.

3.3.1.2 Roulette wheel selection

In roulette wheel selection, the circular wheel is divided into segments representing individuals. Where each segment size is proportional to the fitness of the individual, as represented in Figure 3.2. The wheel is rotated and a specific point on the perimeter of the wheel is selected as indicated in the figure. The segments that comes in front of that specific point is chosen

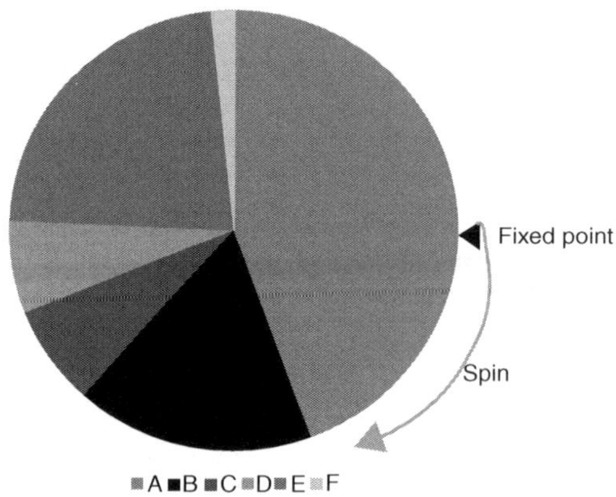

Chromosome	Fitness value
A	8.2
B	3.2
C	1.4
D	1.2
E	4.2
F	0.3

Figure 3.2 Roulette wheel selection diagram.

as a parent [17]. Similar process is followed for choosing the next parent. When the wheel is revolved, an individual with better fitness will have more probability of stopping in front of the designated point, as it has the larger segment on the wheel. Hence, likelihood of selecting an individual is directly related to his fitness value.

3.3.1.3 Ranking selection

In rank selection, the population is first ranked, and then each chromosome's fitness is assigned based on its position in that ranking. The worst chromosome receives a fitness value of 1, the second-worst gets a fitness of 2, and so forth, with the best chromosome receiving a fitness value equal to N, where N is the total number of chromosomes in the population [18]. Following that, all chromosomes have a chance to be chosen. Premature convergence can be avoided using rank-based selection techniques. However, because it sorts the populations based on fitness value, it can be computationally expensive. However, because the best chromosomes do not differ as much from the others, this strategy may result in slower convergence.

3.3.2 Crossover

Crossover is a genetic characteristic that involves the joining of two chromosomes (also known as parents) to form a new chromosome (also called offspring). If the new chromosome inherits the best qualities from both parents, the result of crossover may be better than both parents [19]

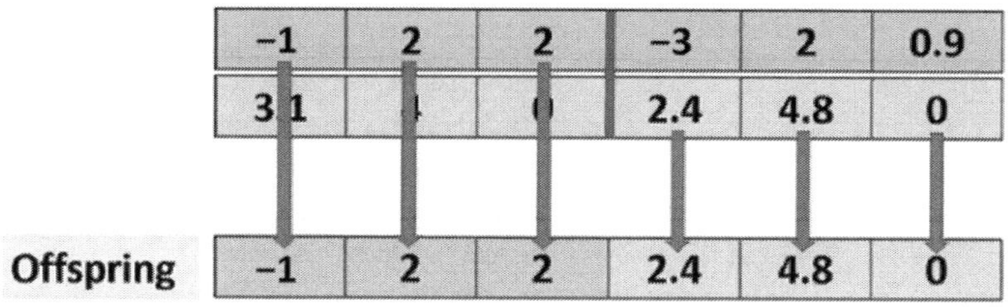

Figure 3.3 Crossover.

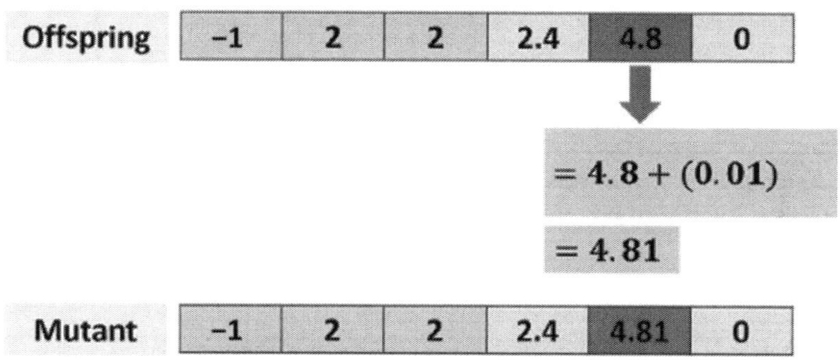

Figure 3.4 Mutation.

as shown in Figure 3.3. During evolution, crossover happens according to a user-defined crossover probability. The new offspring will inherit certain characteristics from one parent and others from the other. Assume first parent is 10001011 and second parent is 10011111; after completing the crossover, the outcome can be 10011111, which has some part of the first parent and some part of the second parent.

3.3.3 Mutation

After the crossover is completed, mutation may occur. Which is done to avoid all population solutions falling into a local best of the current problem. The encoding as well as the crossing influences the mutation. When encoding permutations, for example, the mutation could mean switching two genes [20]. The new offspring are randomly altered by mutation as shown in Figure 3.4. We can encode binary data by altering a few randomly selected bits from 1 to 0 or 0 to 1. The mutations for the parent 10001011 could be 10001111 and for the parent 10011111 could be 10010111.

3.3.4 Formulation of objective function and fitness function

The objective function is used to determine a solution's worth. The output of this objective function is then turned into a fitness value that is utilized to determine the quality of a solution. Because GA is an optimization tool, the objective function is often a maximization or minimization function that includes all variables. The fitness function was designed to create a bimodal histogram in the grey level intensities, allowing for automated image processing steps to evaluate the quality of the individuals throughout the optimization. Individuals with a histogram with two well-separated normal distributions and an equal distance from the optimal threshold are favored by the fitness function we utilize.

GAs have numerous benefits over classical optimization algorithms. These algorithms are very versatile and are capable of handling a wide range of objective functions. The objective function may be stable or dynamic, linear or nonlinear, continuous or discontinuous, or even affected by random noise. The ability of GAs to tackle complex problems and their inherent their inherent parallelism are the two most notable advantages. Because multiple offspring in a population operate independently, the population (or any subset) can hunt in multiple directions at the same time. This capability of GAs makes the parallelization of algorithms very easy, and it also simplifies the modification of different parameters and even groups of encoded strings. However, GAs have some drawbacks as well. The selection criteria for the new population, the design of objective function, the choice of population size, and the selection of key factors like mutation and crossover rates must be handled with utmost care. Any misstep in the choices of these parameters can make it difficult for the algorithm to converge optimally and resulting in ineffective outcomes. Despite these limitations, GAs are nevertheless extensively used nonlinear optimization strategies today.

As a result, GAs is being used extensively in the field of engineering over the past decade. GAs are now being seen as the powerful tool for design optimization. This research work explores the different applications of evolutionary algorithms to solve complex optimization problems, challenges, and managing uncertainty in typical industrial scenarios. Here, the GA is utilized as an optimization mechanism in the supply chain planning process under uncertainty to determine the best production, inventory level, and distribution strategies.

3.4 METHODOLOGY

This research work aims to create a framework for creating the finest possible images by using an algorithm that considers all restrictions to produce better outcomes. For accurate and efficient picture enhancement, it employs

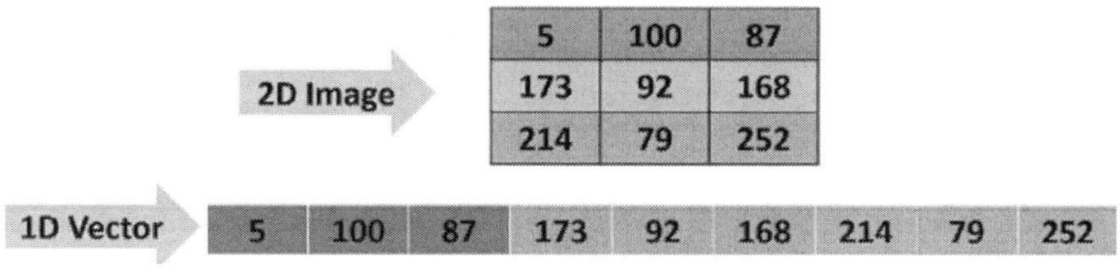

Figure 3.5 Data representation of input image.

the GA. It takes a picture as input and then enhances it. We start by interpreting an image and setting up the fitness feature. The GA begins with a randomly produced image that is similar in shape to the input image. This randomly generated image is continuously evolved utilizing crossover and mutation, as well as GA, until it achieves a better solution than the original.

3.4.1 Data representation

The initial step in solving a GA optimization problem is to consider the best way to describe the data. The chromosome (i.e. solution) is accepted as a 1D row vector by GA, but the input image is not one-dimensional. If the image is binary or grayscale, it may be 2D. If the input image is color, there may be more than two dimensions. If the color is RGB, there are three dimensions, one for each channel. A 1D vector must be used to represent data with more than one dimension. How can we turn multidimensional (MD) data into a 1D vector?

Starting with the simplest scenario, transforming a 2D image (or 2D matrix) into a 1D vector necessitates merging the two dimensions into a single dimension as shown in Figure 3.5. The matrix has numerous rows, and all of these rows must be merged into a single row. This is accomplished by stacking the several rows together. The following diagram depicts this. The illustration depicts a three-row, three-column image/matrix. There are $3 \times 3 = 9$ elements in all.

The vector length will undoubtedly be 9 when converted to a 1D vector. The first three elements of this vector will be drawn from the image's first row. The vector's next three items will come from the second row, and the vector's last three elements will come from the last row.

3.4.2 Fitness function

The fitness or objective function is a key segment of the GA, as it determines the quality of each image iteration based on a set of quantitative indices. As greyness and spatial ambiguity measurements for image enhancement and segmentation challenges, the proposed technique uses entropy (H), compactness (COMP), index of area coverage (IOAC), and their combinations as quantitative indices for evaluating picture quality. The average amount of fuzziness in the greyness of an image (X) is the degree of difficulty (ambiguity) in identifying whether a pixel should be considered black (dark) or white (bright) based on global information. Compactness and IOAC take local information into consideration and indicate the degree of fuzziness in an image's shape and geometry (spatial domain). As a result, the concept of minimizing these ambiguous measurements can be used to build a fitness (assessment) function.

During each iteration, the fitness function calculates a composite score by applying a weighted formula that integrates these indices. Higher scores indicate better-defined image features and clearer contrasts, guiding the GA to select and breed images that move closer to an optimal, high-quality output. By evolving images towards a bimodal histogram with distinct peaks in intensity, the fitness function drives the enhancement process, ensuring the GA consistently favors improved clarity and contrast.

3.5 EXPERIMENTATION AND RESULTS

GA is utilized in a range of complicated problems; a second level job must be used to regulate their parameters for efficiency's sake. As previously stated, the rate of mutation and crossover is known to have a significant impact on the efficiency and behavior of GA. Some researchers utilize tuning of the different parameters to select them, such that it identifies the best value of the parameters before the execution and then run the GA with these. Alternatively, other researchers have proved that utilizing control parameters (deterministic, adaptive) is more efficient in generating best or near-best results.

According to certain studies, high crossover rates of 80–95% and modest mutation rates of (0.1–1) should be used. The researcher advises utilizing a high mutation e.g. 50% with crossover and a small population in another study. Without any feedback from the search process or user involvement, the deterministic control parameter alters the parameter value during the execution. As a result, this kind of feature setting should be utilized to try to figure out which behavior the parameter should deal with in order to be effective and to discover the best or near-best answer. Using high crossover rates with low mutation rates, along with a large population size, promotes diversity within the population. However, in smaller populations, higher mutation rates are employed to introduce diversity and enhance search quality. By adjusting crossover and mutation rates, GA efficiency can be improved through a balanced approach to population diversity.

3.5.1 Qualitative and quantitative results

Figures 3.6 and 3.7 show the input image and the enhanced image generated after applying the enhancement process using different numbers of generations and different crossover rate. The similar approach can also work on images other than those are related to medical field, as shown in Figures 3.8 and 3.9. As shown in these figures, increasing the number of generations and adjusting crossover rates significantly impacted the quality

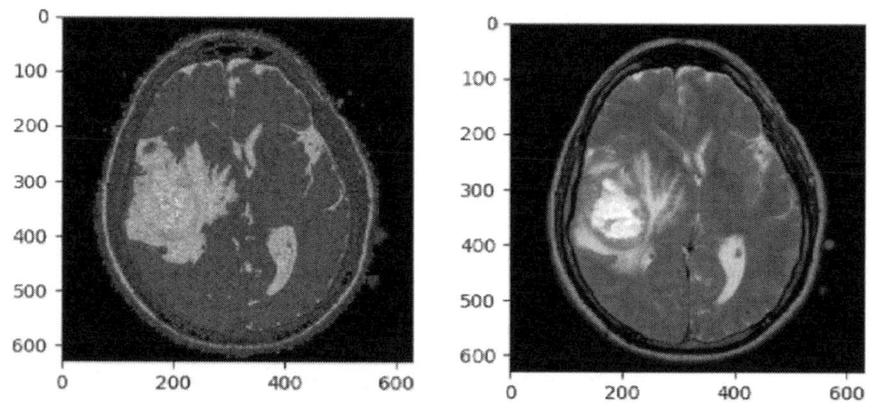

Figure 3.6 Input image versus enhanced image: generations 500, crossover rate 0.7, and mutation rate 1%.

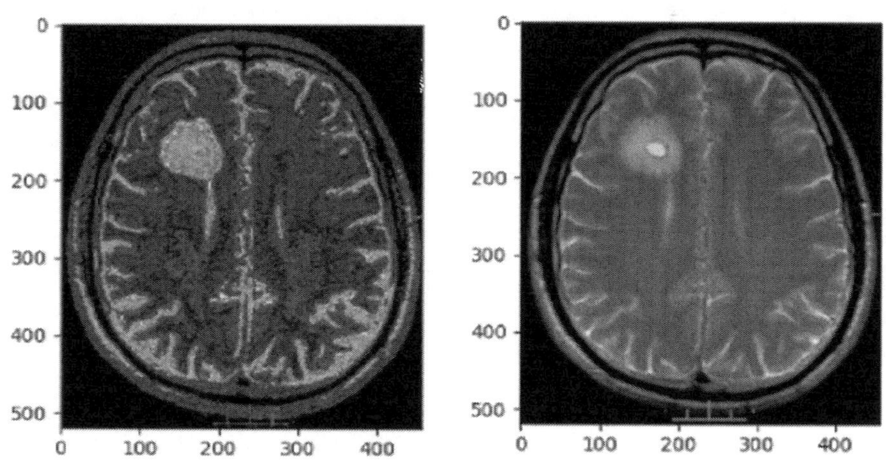

Figure 3.7 Input image versus enhanced image: generations 600, crossover rate 0.4, and mutation rate 1%.

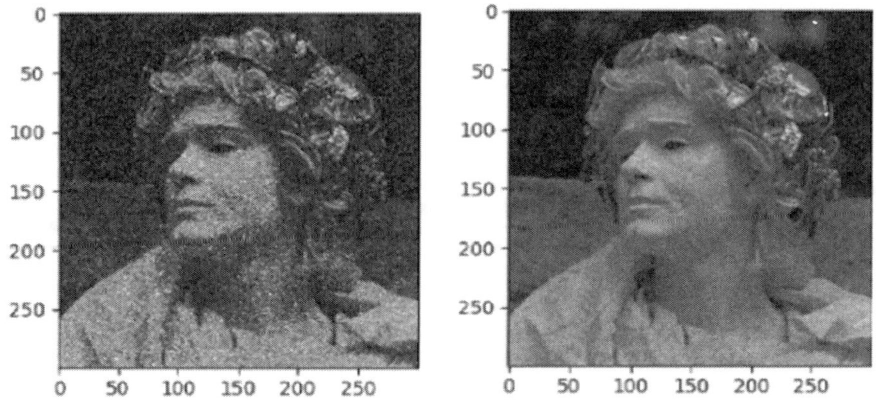

Figure 3.8 Input image versus enhanced image: generations 500, crossover rate 0.3, and mutation rate 1%.

Figure 3.9 Input image versus enhanced image: generations 500, crossover rate 0.5, and mutation rate 1%.

of the enhanced images. With an optimal combination of 500 generations, a 0.4 crossover rate, and a 1% mutation rate, the images achieved the highest clarity and contrast enhancement. These optimized parameters resulted in images where critical medical features were highlighted, allowing for easier anomaly detection in clinical settings.

3.5.2 Time efficiency and scalability

To assess scalability, tests were conducted on image datasets of various resolutions. While the GA performed efficiently for smaller image sets, higher-resolution images required significantly more processing time, underscoring a limitation of the framework in terms of computational cost. Further optimization or parallel processing could help address this limitation, making the GA framework more viable for large-scale applications.

3.6 LIMITATIONS

One of the main limitations of the proposed GA framework is the high-computational cost associated with iteratively evolving large datasets. Each generation requires a fitness evaluation, crossover, and mutation processes that can be computationally intensive, especially for high-resolution medical images. As the dataset size increases, so does the number of generations and operations required to converge on an optimal solution. This prolonged processing time may limit real-time applications, making the framework less suitable for scenarios where immediate image enhancement is essential.

Moreover, the GA's reliance on numerous parameters—such as mutation rate, crossover rate, and population size—demands careful tuning to maintain efficiency, which can add further to the processing time if multiple configurations need to be tested. Future work could explore optimized parallel processing techniques or hybrid models that combine GAs with other, less resource-intensive algorithms to improve efficiency without compromising image quality.

3.7 CONCLUSION AND FUTURE WORK

This work has proven to be an extremely important step for our project. This research has the potential to serve as a prototype for a wide variety of medical specialties across many different dimensions. It is possible to improve upon and retrieve with higher quality various ancient and old photographs that are typically blurry and a little bit on the lower side in terms of what is considered to be their overall quality. These images can then be stored for the future. The ability to learn about and work on the process of image enhancement utilizing cutting-edge techniques like GAs was one of the many things that I picked up from working on this project. The project can be developed further and put into use in the future by adding a few more features and attributes to it. These additions will make the project more user-friendly by making it more convenient and less difficult to use. In future, the integration of all the tools into a single web app or desktop program so that work can be done without interruption. Make an effort to develop additional application tools so that the process of improvement can be completed more quickly. Improve the output by enhancing the performance of the currently used method.

REFERENCES

[1] Z. Li *et al.*, "TFCNs: A CNN-transformer hybrid network for medical image segmentation," *Lect. Notes Comput. Sci.*, vol. 13532, pp. 781–792, 2022. doi:10.1007/978-3-031-15937-4_65/TABLES/5

[2] A. Draa and A. Bouaziz, "An artificial bee colony algorithm for image contrast enhancement," *Swarm Evol. Comput.*, vol. 16, pp. 69–84, Jun. 2014. doi:10.1016/j.swevo.2014.01.003

[3] K. Qin, K. Xu, F. Liu, and D. Li, "Image segmentation based on histogram analysis utilizing the cloud model," *Comput. Math. Appl.*, vol. 62, no. 7, pp. 2824–2833, Oct. 2011. doi:10.1016/J.CAMWA.2011.07.048

[4] L. Cui *et al.*, "DDSE: A novel evolutionary algorithm based on degree-descending search strategy for influence maximization in social networks," *J. Netw. Comput. Appl.*, vol. 103, pp. 119–130, Feb. 2018. doi:10.1016/j.jnca.2017.12.003

[5] X. S. Yang, *Engineering Optimization: An Introduction with Metaheuristic Applications*. John Wiley, 2010. doi:10.1002/9780470640425

[6] S. N. Kumar, A. Lenin Fred, H. Ajay Kumar, and P. Sebastin Varghese, "Firefly optimization based improved fuzzy clustering for CT/MR image segmentation," in *Intelligent Systems Reference Library*, Springer, 2019, pp. 1–28. doi:10.1007/978-3-319-96002-9_1

[7] R. Rajabioun, "Cuckoo optimization algorithm," *Appl. Soft Comput. J.*, vol. 11, no. 8, pp. 5508–5518, 2011. doi:10.1016/j.asoc.2011.05.008

[8] D. Karaboga and B. Basturk, "Artificial bee colony (ABC) optimization algorithm for solving constrained optimization problems," in *Lecture Notes in Computer Science (including subseries Lecture Notes in Artificial Intelligence and Lecture Notes in Bioinformatics)*, Springer, 2007, pp. 789–798. doi:10.1007/978-3-540-72950-1_77

[9] Q. Jiang, G. Song, G. Cong, Y. Wang, W. Si, and K. Xie, "Simulated annealing based influence maximization in social networks," *Proc. Natl. Conf. Artif. Intell.*, vol. 1, pp. 127–132, 2011. doi:10.1609/aaai.v25i1.7838

[10] K. Aggarwal and A. Arora, "Influence maximization for MOOC learners using BAT optimization algorithm," *Int. J. Fuzzy Syst. Appl.*, vol. 11, no. 2, pp. 1–19, Jan. 2022. doi:10.4018/ijfsa.296588

[11] G. Beni and J. Wang, "Swarm intelligence in cellular robotic systems," *Robot. Biol. Syst. Towar.*, vol. 102, pp. 703–712, 1993. doi:10.1007/978-3-642-58069-7_38

[12] S. Mirjalili, S. M. Mirjalili, and A. Lewis, "Grey wolf optimizer," *Adv. Eng. Softw.*, vol. 69, pp. 46–61, Mar. 2014. doi:10.1016/J.ADVENGSOFT.2013.12.007

[13] D. Karaboga, "An idea based on honey bee swarm for numerical optimization," *Tech. Rep. TR06, Erciyes Univ.*, no. TR06, p. 10, 2005,

Accessed: Aug. 30, 2024. [Online]. Available: https://abc.erciyes.edu.tr/pub/tr06_2005.pdf

[14] J. Kennedy and R. Eberhart, "Particle swarm optimization," *Proc. ICNN'95 – Int. Conf. Neural Networks*, vol. 4, pp. 1942–1948. doi:10.1109/ICNN.1995.488968

[15] J. H. Holland, "Genetic algorithms understand genetic algorithms," *Sci. Am.*, vol. 267, no. 1, pp. 66–73, 1992. doi:10.2307/24939139

[16] O. Stitini, S. Kaloun, and O. Bencharef, "An improved recommender system solution to mitigate the over-specialization problem using genetic algorithms," *Electronics*, vol. 11, no. 2, p. 242, Jan. 2022. doi:10.3390/electronics11020242

[17] H. Saruhan, "Design optimization of rotor-bearing systems using genetic algorithms.pdf," 2001. Accessed: Aug. 31, 2024. [Online]. Available: www.proquest.com/openview/552f782a635144207e5fee770d2cbb0e/1?cbl=18750&diss=y&pq-origsite=gscholar&parentSessionId=7IdHq6LdEeF%2BMV7U%2BTQnTFRtbeMNj8gKetmG3SZL9Q4%3D

[18] L. Haldurai, T. Madhubala, and R. Rajalakshmi, "A study on genetic algorithm and its applications," *Int. J. Comput. Sci. Eng.*, vol. 4, no. 10, pp. 139–143, 2016.

[19] F. A. Zainuddin and M. F. A. Samad, "A review of crossover methods and problem representation of genetic algorithm in recent engineering applications," *Int. J. Adv. Sci. Technol.*, vol. 29, no. 6, pp. 759–769, 2020, Available: www.researchgate.net/publication/341930670

[20] M. Srinivas and L. M. Patnaik, "Adaptive probabilities of crossover and mutation in genetic algorithms," *IEEE Trans. Syst. Man Cybern.*, vol. 24, no. 4, pp. 656–667, 1994. doi:10.1109/21.286385

Chapter 4

Machine learning security on drones or UAV

Meenal Jain and Kirti Aggarwal

4.1 INTRODUCTION

4.1.1 Working of drones

4.1.1.1 Network infrastructure of drones

To enable successful communication, control and data transmission, unmanned aerial vehicles (UAVs) or drones rely on a range of network infrastructure components. Understanding these infrastructural components is critical for understanding the intricacies of drone operations:

1. **Communication Links:**
    - Radio Frequency (RF) Links: Drones often use RF links, such as 2.4 GHz and 5.8 GHz, are common for control and telemetry.
    - Wi-Fi: Consumer drones frequently utilise Wi-Fi connections for communication between the drone and a remote controller or smartphone app.
    - Cellular Networks: Commercial and enterprise-level drones are equipped with cellular connectivity, allowing for extended-range communication, data transfer and live video streaming.
    - Satellite Communication: In remote or offshore areas, drones may rely on satellite communication to maintain connectivity, ensuring global coverage.
2. **Data Links:** Drones provide a variety of data kinds, such as photos, movies, sensor readings and telemetry data. Data connections, both cable and wireless, are used to transport data between the drone and ground systems.
3. **GPS Receivers:** Most drones include GPS receivers, which allow them to detect their specific location and assure correct navigation. GPS data is an essential component of drone network architecture, allowing for safe and efficient flying.

 DOI: 10.1201/9781003569619-4

4. **Telemetry Systems:** Telemetry systems send real-time data about the drone's status to the ground control station, including altitude, speed, battery level and sensor readings. This information is critical for tracking and managing the drone's performance.
5. **Remote Controllers:** Operators utilise remote controllers or transmitters to convey control commands to the drone. These controllers communicate with the drone via RF or Wi-Fi connectivity.

These numerous components compose the network backbone of drones, allowing them to perform efficiently in a variety of applications. The infrastructural elements chosen are determined by considerations such as the drone's function, range requirements, operational environment and regulatory restraints. Understanding this infrastructure is critical for understanding the security issues and potential vulnerabilities that may be addressed in the context of drone operations.

Drones are susceptible to a variety of hostile assaults, such as data eavesdropping, GPS spoofing, radio frequency jamming and unauthorised access. This emphasises the urgent need for extensive security measures to safeguard both the drones and the private information they carry. Drone network architecture often incorporates security features including encryption, authentication, intrusion detection systems and network monitoring to prevent unauthorised access and data eavesdropping.

4.1.2 Adversarial attacks

A wide variety of algorithms have developed in the complex field of machine learning (ML) security, each one specifically designed to look for and take advantage of flaws in models. These algorithms primarily focus on the exciting field of adversarial attacks, which explores and manipulates the very underpinnings of ML models.

1. **L-BFGS Algorithm:** The L-BFGS method is an aggressive optimisation approach. It is a white-box attack approach. L-BFGS generates adversarial samples by repeatedly improving perturbations on input data in order to optimise the model's loss. This approach is successful, although it is computationally demanding.
2. **Fast Gradient Sign Method (FGSM):** FGSM is a white-box attack algorithm that is both quick and efficient. It modifies the input data by performing a single step in the direction of the loss function's gradient with respect to the input. Although FGSM produces adversarial cases fast, it does not always achieve high success rates.
3. **Basic Iterative Method (BIM) and Projected Gradient Descent (PGD):** BIM and PGD are white-box iterative attack strategies. They create

adversarial samples by doing numerous iterations of perturbation on input data. BIM begins with a modest perturbation and adds noise repeatedly to maximise the model's loss. PGD is a more powerful form of BIM that projects perturbations onto a ball to keep them within a restricted range.

4. **Momentum Iterative Attack:** The motion iterative method is a white-box iterative method that improves BIM by adding motion. It uses gradient information accumulated over iterations to make greater modifications to perturbations. This method aids in avoiding local optima and frequently yields stronger adversarial samples.
5. **GPS Spoofing:** GPS spoofing is the sending of fake or altered GPS signals in order to fool GPS receivers (see Figure 4.1). The purpose of GPS spoofing is to trick the target device or system into thinking it is somewhere else than it is. GPS spoofing is especially troublesome in the case of drones or UAVs. Drones frequently rely on GPS for precise navigation, waypoint tracking and flying stability. When a drone is subjected to GPS spoofing, it can be misled into taking erroneous flight paths, possibly leading to dangerous scenarios. It can also jeopardise the security of operations that rely on precise location data.

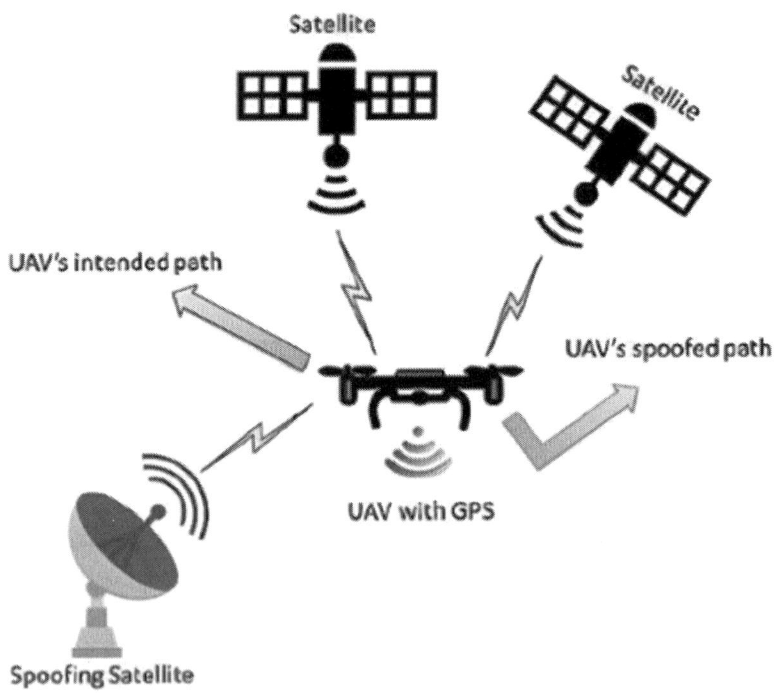

Figure 4.1 GPS spoofing.

6. **Distributionally Adversarial Attack:** Individual samples are not targeted by distributionally adversarial assaults, which focus on perturbing data distributions. They strive to reduce the statistical divergence between the original and altered data distributions. These attacks evaluate model resilience by taking into account the complete data distribution and frequently demonstrate model transferability.
7. **Blockchain-based attack:** A blockchain-based attack on a drone could take a number of different forms. For example, an attacker modify the drone's data and disrupt the drone's communication with its ground control station.

In conclusion, adversarial attack strategies are an essential and dynamic part of the ongoing endeavour to enhance the security of ML models. By actively probing and pushing these models with hostile inputs, security practitioners and researchers may uncover weaknesses, evaluate threats realistically and promote continuous development in the robustness and resilience of AI systems. Adversarial attack techniques are becoming more and more vital in preserving the reliability and integrity of AI-powered systems as ML expands into more critical industries.

4.2 LITERATURE SURVEY

Following a close review of the aforementioned research publications discussed in Table 4.1, it is clear that the multiple applications of UAVs in the military, public and technological sectors have resulted in significant advancement in a variety of disciplines. Advances in cybersecurity frameworks, safe autonomous navigation and trajectory planning-focused navigational systems, and secure communication for the Internet of Drones (IoD) are among them. The incorporation of satellite, 5G and UAV technologies into telecommunication networks has highlighted the importance of cybersecurity measures to avoid cyberattacks. ML techniques are utilised to improve security in UAV networks, specifically to distinguish between hostile and benign network traffic. An emphasis has also been placed on overcoming adversarial attack vulnerabilities in deep learning (DL) algorithms, as well as assessing the advantages and risks connected with the growing use of drones in armed services, healthcare and agricultural applications.

Research examines how UAV technology interacts with the Internet of Things (IoT) and how this may affect smart cities in the future. The widespread use of drones in these metropolitan environments presents significant risks to public safety, privacy and cybersecurity. The increased use of drones has spurred development on secure data gathering and storage systems such as SecureIoD, which uses blockchain-based storage methods, secure key establishment and authentication to protect IoD data. Although difficulties with managing massive data quantities still exist, the integration of UAVs

Table 4.1 Key contributions in UAV security and applications

Author	*Summary*
Samanth et al. [1]	UAVs are used in both armed-services and public contexts. While armed-services applications include border surveillance and intruder reporting to Ground Control Stations (GCS), public usage include construction site and catastrophe monitoring. Proposed UAV frameworks are evaluated using open-source UAV simulators such as AirSim. The Internet of Drones (IoD) requires safe communication; studies of secure IoD communication frameworks use notions from cryptography. NS3 and OMNeT++ are popular network simulators used for this kind of study
Vattapparamban et al. [2]	Drones are predicted to play a significant part in charge of maintaining smart city security and monitoring, transporting commodities and merchandise and acting as mobile hotspots for broadband wireless access. Drone usage for future smart cities will, however, also inevitably raise a number of technological and social issues that require attention. This survey article aims to analyse several drone-related topics related to cybersecurity and privacy. We will also share representative findings about drone-based cyberattacks
Shrestha et al. [3]	Recent increases in the use of UAVs in everything from parcel delivery to wildlife protection are driving the advancement of telecommunication networks to incorporate 5G, satellite and UAV technology. But because UAV nodes are more and more vulnerable to hacks, security worries are developing. This study presents a machine learning-based 5G network security model that utilises UAVs and satellites to identify seven modern attack types. In order to improve security, the article shows how well ML systems can differentiate between malicious and benign UAV network packets
Ren et al. [4]	Algorithm security is crucial in the quickly developing fields of deep learning (DL) and artificial intelligence (AI). Much attention has been drawn to the susceptibility of DL algorithms to adversarial samples, which may fool human observers while inducing misbehaviour in models. Adversarial attacks have generated attention in the machine learning and security fields as they have shown to be feasible in real-world situations. The theoretical underpinnings, algorithms and practical uses of adversarial attack strategies are presented in this work. It also includes research initiatives in the sphere of defensive tactics. The paper's conclusion highlights unresolved issues to promote more study in this important field
Emimi et al. [5]	In healthcare, drones deliver critical medical supplies to remote areas during emergencies. In the armed-services, they enhance security and surveillance efforts. However, drones also pose risks like injuries and property damage, especially when mishandled or experiencing component failures. Extremists can potentially hijack drones for their purposes. This paper comprehensively analyses agricultural, medical and armed-services drones, examining their strengths, weaknesses and opportunities

Table 4.1 (Cont.)

Author	*Summary*
Lu et al. [6]	DNNs are often employed in UAV visual navigation and recognition; yet, they are susceptible to real-time picture perturbations and adversarial camouflage. This research proposes two ensemble techniques for proactive and reactive adversarial defence on UAVs utilising CNN and transformer models. Using adversarial transferability, the proactive method mixes output probabilities from basic DNN models. The reactive method combines average output confidence scores and hidden features with adversarial detectors' scoring systems. Their usefulness in improving the resilience of recognition tasks is demonstrated by experiments conducted on optical and synthetic aperture radar (SAR) datasets. Additionally, AREP-RSIs, a framework for assessing adversarial resilience in remote sensing models, is introduced in this study
Shankar et al. [7]	One revolutionary technique in the Internet of Medical Things (IoMT) is health monitoring. Through the connection of sensors and actuators, data may be captured, monitored, stored and displayed through mobile or online apps. Vital sign monitoring, or IoMT, assists healthcare providers in determining a patient's state of health and suggesting appropriate treatment. This research presents a novel UAV-integrated health monitoring system that enables clinicians to view and analyse patient data remotely. This IoMT-based UAV system lowers readmissions and hospital stays, however processing large amounts of data presents processing issues. The capability to capture and save video for doctor surveillance together with several parameter measures is a special feature
Elmokadem and Savkin [8]	Unmanned aerial vehicles (UAVs) have rapidly evolved, becoming valuable for various armed-services and public applications, reaching remote areas, saving time and enhancing safety. The focus of UAV development is shifting toward full autonomy, reducing human intervention. This paper surveys recent advancements in UAVs, emphasising safe autonomous navigation, a critical aspect of the system. It delves into advanced methods for 3D avoidance manoeuvres and safe trajectory planning, addressing research challenges in UAV development
Aldaej et al. [9]	The development of drones has created new opportunities in many industries, especially for tiny drones. Through the Internet of Things (IoT), they make navigational services possible. However, due to design problems, drone networks provide privacy and security issues, requiring a secure network. This paper highlights the importance of security while examining current privacy and security issues in drone networks (NoD). For data categorization, a hybrid machine learning method that combines random forest and logistic regression is employed. The suggested strategy improves NoD cybersecurity by using cutting-edge AI-inspired techniques, guaranteeing protection. Improved performance is demonstrated via validation using a difficult dataset in terms of statistical metrics and efficiency

(continued)

Table 4.1 (Cont.)

Author	*Summary*
Meshram and Reddy [10]	The site must be inspected by site owners, architects, structural designers, waterproofing companies, and others for quality control, construction progress and material unloading, among other reasons. Each of these parties needs to visit the site many times in order to do this task, which raises the site's cost. In addition, a lot of people visit the site, which means that construction and management issues arise frequently. This study suggests a ground-breaking drone-based site monitoring and management system as a workaround. In order to find disparities in material unloading, construction quality, site progress, labour activities and other areas, machine learning techniques are employed in this study. The underlying method is flexible and may be used for many types of events for better upkeep of the site quality. Based on statistical analysis conducted at five sites of varying sizes, the drone-based machine learning model may reduce construction costs by over 10% and increase the quality of construction by 15%, therefore increasing the overall efficiency of building sites and development
Medaiyese et al. [11]	Concerns about security and privacy now have a new angle thanks to drones. Few, if any, strict restrictions regulate who is allowed to purchase or own a drone. This means that anyone may use these planes to sneak into areas that are private or restricted. Using a drone detection and identification (DDI) system is one way to find and identify drones in a given region. DDI systems can utilise a range of sensing modalities, including radio frequency (RF) signals, video, sounds and thermal imaging, to identify an invading drone. In this research, we saw an RF-based machine learning DDI system based on low band RF data from the drone-to-flight controller link. We develop three machine learning models that detect and identify drone presence, kind and operation mode using the XGBoost algorithm. These three XGBoost models were analysed using 10-fold cross validation, and the average accuracy that we were able to obtain was 99.96%, 90.73%, and 70.09%, respectively
Sachdeva et al. [12]	Unmanned aerial vehicles (UAVs) are susceptible to various cyberattacks and vulnerabilities. Additionally, they can be seized and utilised by illicit organisations, leading to cybercrime. These vulnerabilities can be used by adversaries to cause data loss, property damage and even fatalities. While keeping an eye on UAV activities can assist in identifying attacks like GPS spoofing, jamming, false information dissemination and blackhole and grey hole development, it may not allay privacy concerns. This paper explains how UAVs may communicate securely using blockchain technology. Our approach involves building a trustworthy and secure UAV ad hoc network as well as smart contracts. This network will be secure from malevolent attacks and resilient to a wide range of network hazards

into the Internet of Medical Things (IoMT) has produced creative health monitoring systems that enable medical practitioners to access and analyse data remotely.

More autonomy and less human involvement are the goals of UAV technological advancements, with a focus on safe navigation methods. Strong cybersecurity standards are required in light of privacy and security vulnerabilities in drone networks. The use of ML techniques, like decision tree and logistic regression, has shown improved cybersecurity protocols for drone networks. Furthermore, a number of studies highlight how artificial intelligence (AI) and ML are revolutionising a number of drone-related fields, including intrusion detection systems, site monitoring and communication security.

Blockchain technologies, 5G networks and AI all work together to create potentially intelligent and safe drone communication systems as shown in Figure 4.2. Additionally, with an emphasis on security in smart cities, AI-driven 5G apps provide cutting-edge capabilities in virtual emotion systems. Blockchain-based solutions show promise in protecting networks from criminal attacks and network threats while also safeguarding communication in UAV systems. As this is going on, intelligent intrusion detection systems are being developed to combat common threats to UAVs. These systems use techniques such as PCA and single-class classifiers to identify

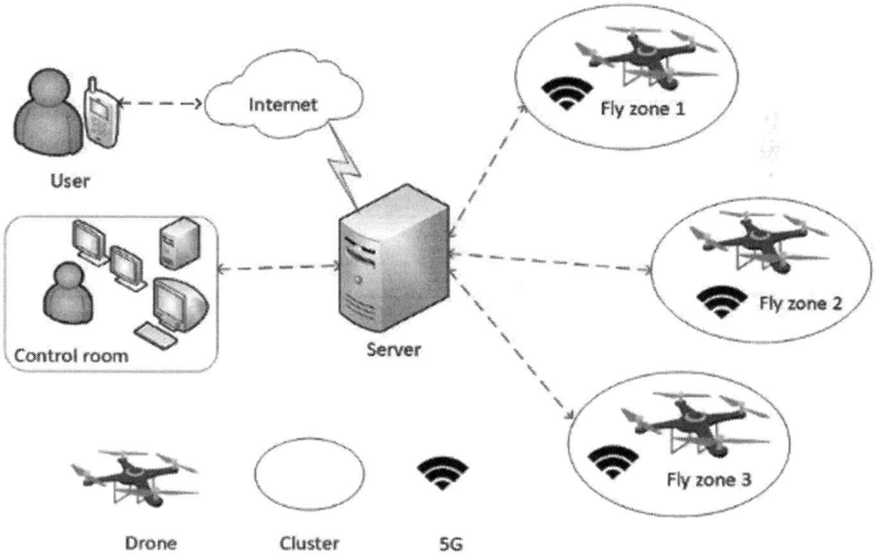

Figure 4.2 Drone communication over 5G network.

and neutralise attacks like jamming and spoofing, protecting UAVs even in situations where communication is disrupted.

4.3 COMPARATIVE ANALYSIS BASED ON FOCUS, METHODOLOGY AND RESULTS OF REFERENCED PAPERS

4.3.1 Machine learning-based approaches

'Machine-Learning-Enabled Intrusion Detection System for Cellular Connected UAV Networks' security model creation for 5G and satellite-based UAV networks, employing ML for packet distinction and attack identification. The research examines the consequences of combining UAVs, 5G networks and satellite technologies, with an emphasis on the recent boom in UAV deployment across several sectors. It tackles the growing security issues with UAV nodes, which are vulnerable to assaults because to inherent weaknesses and increasing data volumes. The study offers unique UAVs using the 5G network security model that utilise ML approaches to identify and categorise several forms of modern attacks on UAV networks. Comparative evaluations are carried out to analyse the performance and accuracy of ML algorithms, with a focus on the decision tree algorithm's usefulness in identifying different UAV network threats [3].

'Adversarial Attacks and Defences in Deep Learning' examines vulnerabilities of DL algorithms to adversarial attacks in UAVs, discussing defence techniques.

This study focuses on the susceptibility of DL algorithms used in UAVs to adversarial assaults. It dives into the theoretical underpinnings, methods and real-world applications of adversarial attack strategies on DL algorithms.

The paper provides a thorough review of protection mechanisms in response to adversarial attacks, encompassing the domains of ML and security research [4].

'Adversarial Robustness Enhancement of UAV-Oriented Automatic Image Recognition Based on Deep Ensemble Models' uses CNN and transformer models, a paper that describes research conducted on optical and synthetic aperture radar (SAR) datasets.

Using transformer models and convolutional neural networks (CNNs), the study suggests and assesses two ensemble methods for proactive and reactive adversarial defensive mechanisms in UAVs. Whereas the reactive technique combines adversarial detector scoring functions with average output confidence ratings and hidden features, the proactive method combines adversarial detector scoring functions with average output confidence scores and hidden features.

Optical and SAR datasets are used in experiments to test the defensive mechanisms' ability to improve recognition task performance, especially in hostile environments [6].

'Smart IoMT Framework for Supporting UAV Systems with AI' integrates health monitoring with UAVs but lacking specific methodology details and introduces a novel Health Monitoring System connected with UAVs, allowing medical practitioners to view and analyse data remotely. This UAV system based on IoMT intends to minimise hospital stays and readmissions [8].

'Smart cybersecurity framework for IoT-empowered drones: machine learning perspective' shows the technique focuses on the privacy and security vulnerabilities that exist in drone networks (NoD) and emphasises the importance of developing a secure network architecture. The paper dives into current privacy and security problems raised by drone networks, attributing them to network design defects. To address these concerns, the study utilises a hybrid ML approach for successful data categorization that combines logistic regression and random forest algorithms. The technique incorporates sophisticated AI-inspired approaches, which contribute to the enhancement of cybersecurity measures in NoD networks. Data collecting from difficult datasets related to NoD privacy and security problems is part of the study process. The validation method results in an evaluation of the suggested approach's efficiency and statistical performance metrics, demonstrating improvements in NoD network cybersecurity and overall protection [10].

'Machine Learning Applications in Internet-of-Drones: Systematic Review, Recent Deployments, and Open Issues' discusses the broad utility of drones/UAVs and their adaptability in various fields. This research looks into the application of ML in the context of the IoD and UAVs. It examines IoD/UAV ideas, ML applications, current breakthroughs, existing approaches and future research prospects in depth. CNNs are frequently used in IoD-UAV applications, with resource and mobility management gaining the greatest focus, according to the findings. Python is the most popular programming language and publishing activity peaked in 2021. The study emphasises the different potential uses for drones across several industries, emphasising their versatility and practicality in a variety of circumstances. This shows the wide range of applications for drones [13].

'Development of a Machine Learning-Based Drone System for Management of Construction Sites' proposes a drone-based site monitoring system to enhance site quality and reduce construction costs. ML algorithms are used to process site monitoring data collected by drones. To detect abnormalities in material unloading, construction quality and labour activities, these systems use anomaly detection and classification techniques. The solution employs a scalable event-based architecture for site quality management on an ongoing basis. It is anticipated that the use of ML algorithms for site monitoring would result in a significant 10% reduction in building costs and a 15% improvement in construction quality. This enhancement shows how effective drone site management can be [10].

'Machine Learning Framework for RF-Based Drone Detection and Identification System,' focuses on ML-based drone detection and identification solutions to solve privacy and security concerns. Using low band radio frequency data from drone-to-flight controller communications, it applies ML models for drone identification and detection through RF signal analysis. Ten thorough cross-validations are performed on the models to ensure accuracy.

The use of ML algorithms for site monitoring is predicted to significantly cut construction costs by more than 10% while improving building quality by 15%. This advancement demonstrates the effectiveness of deploying drones for site management [11].

'Deep Learning Driven Security in Digital Twins of Drone Network' details a security study using improved LSTM for drone attack prediction, protecting data privacy using differential privacy and testing model performance using Tennessee Eastman process. Cyberphysical system (CPS) data is evaluated using an advanced LSTM network, with a focus on predictive modelling for drone control signal data before an attack. Through the utilisation of the Tennessee Eastman process as a simulation platform to assess the prediction model, the solution employs differential privacy techniques to protect data privacy. An augmented LSTM network paired with differential privacy approaches is expected to increase prediction accuracy and robustness in drone attack prediction. This result indicates improved security in drone systems [14].

'Machine Learning-Based Delay-Aware UAV Detection and Operation Mode Identification over Encrypted Wi-Fi Traffic' describes methods for UAV detection and operating mode identification utilising encrypted Wi-Fi communication characteristics.

Wi-Fi communication metrics like packet size and interarrival time are employed for UAV identification and detection. The method that the system uses for feature selection and prediction performance improvement is called reweighted l1-norm regularisation.

Wi-Fi data may be used to recognise different consumer UAVs with high accuracy (85.7–95.2%) and fast recognition times (0.15–0.35 seconds). This demonstrates the capacity to promptly and precisely detect UAVs [15].

'Securing Internet of Drones Networks Using AI-Envisioned Smart-Contract-Based Blockchain' presents a security architecture for secure communication using blockchain in the IoT/IoDs domain and proposes the use of blockchain technology to provide safe communication between drones in the context of IoT/IoDs. Smart contracts and peer-to-peer cloud servers (P2PCS) enable safe and encrypted data exchanges, while blockchain approaches ensure consensus.

The use of blockchain to secure drone communication is expected to guarantee secure and encrypted communication while protecting data privacy among drones [16].

'Fusion of blockchain and artificial intelligence for secure drone networking underlying 5G communications' proposes a secure drone communication architecture based on blockchain that makes use of 5G and AI technologies.

This article provides a blockchain-based architecture combined with 5G communication networks and AI algorithms for secure drone communication and enhances communication security and network performance by leveraging the InterPlanetary File System (IPFS) as a platform for data storage. 5G networks and AI algorithms are used to provide a dynamic, adaptive, and secure communication infrastructure for drones. This configuration is considered to be beneficial, especially in healthcare contexts [17].

'Research Challenges and Security Threats to AI-Driven 5G Virtual Emotion Applications Using Autonomous Vehicles, Drones, and Smart Devices' discusses AI-based risks in 5G-enabled smart cities and introduces 5G-I-VEmoSYS, a 5G integrated virtual sentiment system powered by AI. AI-based security issues are discussed in 5G-enabled smart cities and offered a 5G integrated virtual sentiment system powered by AI, hinting at AI uses for drone communication monitoring and security. A 5G integrated virtual sentiment system powered by AI for monitoring and safeguarding drone communication in smart city environments are being developed. The goal is to provide dependable and secure communication networks [18].

'Artificial intelligence for intrusion detection systems in Unmanned Aerial Vehicles' introduces a system to detect instrusions (IDS) for UAVs using PCA and single-class classifiers, tackling GPS spoofing and jamming. In an IDS developed to detect GPS spoofing and jamming threats against UAVs, PCA and single-class classifiers are used. This system operates within the UAV's resource-constrained agent devices and sims at high accuracy rates of over 90% in detecting and perhaps neutralising GPS spoofing and jamming assaults against UAVs [19].

4.3.2 Other approaches for security in drones

'Drones for Smart Cities: Issues in Cybersecurity, Privacy, and Public Safety' highlights public safety as key considerations when identifying drone-related issues in future smart cities.

The study report took a progressive stance, projecting how drones will be employed in networked smart cities in the future. It outlines a wide range of possible duties, such as broadband wireless access, delivery services, monitoring and security systems. The report also highlights the technological and social challenges associated with the widespread deployment of drones in future smart cities, notably with regard to public safety, cybersecurity and privacy.

With an emphasis on how drone integration may affect cybersecurity, the survey article creates a comprehensive approach by reviewing and synthesising recent studies, publications and literature on the topic of drone integration into smart cities in the future [2].

'The Current Opportunities and Challenges in Drone Technology' highlights UAV applications in agriculture, healthcare and armed-services sectors. The study of the many applications of drone technology in agriculture, healthcare and armed-services areas. It examines drones' roles and contributions in various industries in depth, highlighting their advantages, potential hazards and potential misuses. The study examines the advantages, disadvantages, possibilities and potential risks provided by drones in agriculture, healthcare and armed-services settings [5].

'SecureIoD: A secure data collection and storage mechanism for internet of drones' explains the creation and testing of SecureIoD, an IoD data collection system that is secure.

Given the limits of drones and the open wireless medium, this study emphasizes the need for data security and privacy in the IoD scenario. SecureIoD, a secure IoD data gathering and storage solution, is introduced in the study. It consists of mutual authentication between drones and zone service providers (ZSPs), secure key creation and blockchain-based data storage.

The miner ZSP is selected via a hybrid proof-of-work and proof-of-stake consensus mechanism. Practical tests and simulations demonstrate that SecureIoD is more secure and effective than previous systems, and security studies support its resistance to attacks [20].

'Towards fully autonomous UAVs: A survey' highlighting recent advancements in UAVs and safe autonomous navigation methods. The article starts by underlining the progress of UAVs across many applications in the armed-services and public sectors, emphasising its worth in remote access, time savings and safety advancements. The research focuses on current UAV advancements, particularly in terms of reaching full autonomy and eliminating human interaction. The study is to assess and evaluate improvements in UAV technology, with a focus on safe autonomous navigation. The process includes a thorough assessment of current literature, technical reports and research papers to accomplish this. Advanced approaches for 3D avoidance manoeuvres and trajectory planning, addressing unique research issues related with UAV development, particularly in the context of safe autonomous navigation [9].

'Privacy and Security Improvement in UAV Network using Blockchain' addresses UAV vulnerabilities and proposes using blockchain to secure UAV communication. Proposes blockchain technology for protecting UAV communication, with an emphasis on smart contracts and the establishment of a secure UAV ad hoc network that is resistant to different network attacks and intrusions. A blockchain-based secure UAV ad hoc network is envisioned to withstand different network attacks and provide a safe and dependable communication environment for UAVs [19].

'Intrusion detection systems in Unmanned Aerial Vehicles' introduces an intrusion detection system for UAVs using PCA and single-class classifiers, tackling GPS spoofing and jamming. In an intrusion detection system developed to discover GPS spoofing and jamming threats against UAVs, PCA and single-class classifiers are used. This system operates within the UAV's resource-constrained agent devices and aims at high accuracy rates of over 90% in detecting and perhaps neutralising GPS spoofing and jamming assaults against UAVs [20].

4.4 CONCLUSION

Throughout our exploration of research in the field of UAVs and drone technology, we have uncovered critical advancements and challenges that shape this innovative domain. UAVs have become indispensable in both military and civilian applications, offering significant improvements in remote accessibility, efficiency and safety. The push toward fully autonomous UAVs represents a major leap forward, aiming to reduce human intervention while prioritizing safe, autonomous navigation. This progress has been examined through techniques for 3D obstacle avoidance and trajectory planning, addressing key challenges in the field. Simultaneously, the rapid expansion of smaller drones has opened new possibilities across various sectors, especially through the integration of the Internet of Things (IoT) for navigation services. However, this growth brings with it serious privacy and security risks within drone networks (NoD), stemming from inherent design vulnerabilities. Addressing these concerns, we explore a novel hybrid ML approach that combines supervised ML techniques, such as decision-based methods and logistic regression, to enhance NoD cybersecurity. This integrated, AI-driven solution shows promise in strengthening the security of drone networks.

REFERENCES

[1] Samanth S, Prema KV, Balachandra M. 2022. Security in internet of drones: a comprehensive review. *Cogent Engineering*. 9(1):2029080. doi:10.1080/23311916.2022.2029080

[2] Vattapparamban E, Güvenç I, Yurekli AI, Akkaya K, Uluağaç S. 2016. Drones for smart cities: Issues in cybersecurity, privacy, and public safety. In: *2016 International Wireless Communications and Mobile Computing Conference (IWCMC)*; Sep 2016; Paphos, Cyprus. IEEE. pp. 216–221. doi:10.1109/IWCMC.2016.7577060

[3] Shrestha R, Omidkar A, Roudi SA, Abbas R, Kim S. 2021. Machine-learning-enabled intrusion detection system for cellular connected UAV networks. *Electronics*. 10(13):1549. doi:10.3390/electronics10131549

[4] Ren K, Zheng T, Qin Z, Liu X. 2020. Adversarial attacks and defenses in deep learning. *Engineering*. 6(3):346–360. doi:10.1016/j.eng.2019.12.015

[5] Emimi M, Khaleel M, Alkrash A. 2023. The current opportunities and challenges in drone technology. *International Journal of Electrical Engineering and Sustainability*. 1(3):74–89.

[6] Lu Z, Sun H, Xu Y. 2023. Adversarial robustness enhancement of UAV-oriented automatic image recognition based on deep ensemble models. *Remote Sensing*. 15(12):3007. doi:10.3390/rs15123007

[7] Shankar N, Nallakaruppan MK, Ravindranath V, Senthilkumar M, Bhagavath BP. 2022. Smart IoMT framework for supporting UAV systems with AI. *Electronics*. 12(1):86. doi:10.3390/electronics12010086

[8] Elmokadem T, Savkin AV. 2021. Towards fully autonomous UAVs: A survey. *Sensors*. 21(18):6223. doi:10.3390/s21186223

[9] Aldaej A, Ahanger TA, Atiquzzaman M, Ullah I, Yousufudin M. 2022. Smart cybersecurity framework for IoT-empowered drones: machine learning perspective. *Sensors*. 22(7):2630. doi:10.3390/s22072630

[10] Meshram K, Reddy NG. 2022. Development of a machine learning-based drone system for management of construction sites. In: *Advances in Sustainable Materials and Resilient Infrastructure*; Springer Singapore. pp. 77–88. doi:10.1007/978-981-16-7074-5_8

[11] Medaiyese OO, Syed A, Lauf AP. 2021. Machine learning framework for RF-based drone detection and identification system. In: *2021 2nd International Conference On Smart Cities, Automation & Intelligent Computing Systems (ICON-SONICS)*; Oct 2021; Yogyakarta, Indonesia. IEEE. pp. 58–64. https://doi.org/10.1109/ICON-SONICS53103.2021.9617168

[12] Sachdeva H, Gupta S, Misra A, Chauhan K, Dave M. 2023. Privacy and security improvement in UAV network using blockchain. *International Journal of Communication Networks and Distributed Systems*. 29(4):383–406. https://doi.org/10.1504/ijcnds.2023.131736

[13] Heidari A, Jafari Navimipour N, Unal M, Zhang G. 2023. Machine learning applications in internet-of-drones: systematic review, recent deployments, and open issues. *ACM Computing Surveys*. 55(12):1–45. https://doi.org/10.1145/3571728

[14] Wu J, Guo J, Lv Z. 2022. Deep learning driven security in digital twins of drone network. In: *ICC 2022-IEEE International Conference on Communications; May 2022; Seoul, Korea*. IEEE. pp. 1–6. https://doi.org/10.1109/ICC45855.2022.9838734

[15] Alipour-Fanid A, Dabaghchian M, Wang N, Wang P, Zhao L, Zeng K. 2019. Machine learning-based delay-aware UAV detection and operation mode identification over encrypted Wi-Fi traffic. *IEEE Transactions on Information Forensics and Security*. 15:2346–2360. https://doi.org/10.1109/TIFS.2019.2959899

[16] Bera B, Wazid M, Das AK, Rodrigues JJ. 2021. Securing internet of drones networks using AI-envisioned smart-contract-based blockchain.

IEEE Internet of Things Magazine. 4(4):68–73. doi:10.1109/IOTM.001.2100044.

[17] Gupta R, Kumari A, Tanwar S. 2021. Fusion of blockchain and artificial intelligence for secure drone networking underlying 5G communications. *Transactions on Emerging Telecommunications Technologies*. 32(1):e4176. https://doi.org/10.1002/ett.4176

[18] Kim, H., Ben-Othman, J., Mokdad, L., Son, J., & Li, C. 2020. Research challenges and security threats to AI-driven 5G virtual emotion applications using autonomous vehicles, drones, and smart devices. *IEEE Network*. 34(6):288–294. doi: 10.1109/MNET.011.2000245

[19] Whelan, J., Almehmadi, A., & El-Khatib, K. 2022. Artificial intelligence for intrusion detection systems in unmanned aerial vehicles. *Computers and Electrical Engineering*. 99, 107784. https://doi.org/10.1016/j.compeleceng.2022.107784

[20] Pu C, Wall A, Ahmed I, Choo KKR. 2022. SecureIoD: A secure data collection and storage mechanism for internet of drones. In: *2022 23rd IEEE International Conference on Mobile Data Management (MDM)*; Jun 2022; Paphos, Cyprus. IEEE. pp. 83–92. https://doi.org/10.1109/MDM55031.2022.00033

Chapter 5

Image forgery detection

Simran Bhardwaj, Kartikey, and Mukta Goyal

5.1 INTRODUCTION

Image forgery detection has emerged as a crucial domain in digital forensics, driven by the widespread use of digital images in various fields, including journalism, legal evidence, and social media. The increasing sophistication of image manipulation techniques, particularly through deep learning-based generative models, has raised significant concerns regarding the authenticity and integrity of digital content.

Deep learning techniques, especially convolutional neural networks (CNNs), have shown remarkable capabilities in identifying forged images by extracting intricate patterns and inconsistencies that may not be visible to the human eye. Recent studies have explored the role of CNN-based models in forgery localization and detection, with approaches leveraging transfer learning to enhance accuracy and robustness [1]. Additionally, advancements in detecting GAN-generated fake images have demonstrated the efficacy of CNN architectures in distinguishing synthetic content from authentic ones [2, 3].

Feature fusion and ensemble CNN models have further improved the reliability of image forgery detection by integrating multiple levels of feature representations to enhance decision-making accuracy [4]. Hybrid deep learning models that combine multiple architectures have also been proposed to create more robust detection frameworks capable of handling diverse types of forgeries, including copy-move and splicing techniques [5]. Furthermore, research has explored deep learning architectures specifically designed for copy-move forgery detection, emphasizing the importance of extracting fine-grained spatial information to improve localization performance [6].

The role of deep feature analysis in copy-move forgery detection has been extensively studied, highlighting the effectiveness of feature-based approaches in identifying duplicated regions within an image [7]. An adaptive deep learning framework has also been introduced to dynamically adjust to different forgery types, improving detection accuracy across various

 DOI: 10.1201/9781003569619-5

datasets [8]. These advancements collectively contribute to the growing field of image forgery detection [3], reinforcing the necessity for continuous innovation to counter emerging threats in digital image manipulation.

The identification and localization of digital picture alterations carried out with the intention of misleading or deceiving viewers constitutes the problem statement of image forgery detection. It has become easier to make phone or modified photographs because of the widespread usage of image editing tools and methods. These images can be used to harm reputations, present false proof, or mislead people. In several industries, such as media, law enforcement, and the court system, where photos are used as evidence, the proliferation of altered images offers considerable issues. Digital images' authenticity and integrity are essential in these situations, and any manipulation can have negative effects. The goal of image forgery detection is to spot any discrepancies or anomalies in digital photographs that could be signs of manipulation, like modifications to the color, texture, or metadata. The difficulty in detecting image forgery stems from the fact that the adjustments might be subtle, and the forger may use a variety of methods to hide them, including steganography.

With the extensive use of digital media, image modification has become an increasingly common occurrence. This makes picture forgery detection novel in its capacity to identify it in digital images. The necessity to guarantee the validity and integrity of digital photographs, particularly in legal and investigative contexts, has encouraged the development of image forgery detection algorithms. The creation of automated methods that can quickly and precisely identify altered images has been one of the key developments in the detection of image forgery. Machine learning algorithms and deep learning models have enabled the detection of subtle alterations in images, uncovering forgeries that might escape even the attention of human experts. The area of image forgery detection spans a wide range of disciplines, including journalism, media forensics, law enforcement, and scientific research. With more and more applications using digital photos, the scope of image forgery detection is growing. When it comes to locating altered photos that could be used as evidence in investigations and judicial procedures, image forgery detection is a critical tool in law enforcement. Digital photos used as evidence must be accurate and authentic; any modification of the images could have serious repercussions. The use of doctorated photos in journalism has the potential to misinform and mislead viewers, harming the credibility and reputation of media organizations. Techniques for detecting picture forgery are used to spot altered photographs and guarantee the veracity of media material. Image forgery detection is used in media forensics to investigate incidents of image modification, such as those involving political propaganda, where images are changed to sway public opinion. Techniques for detecting picture forgeries are used to spot altered photos and offer proof in court.

Image forgery detection is employed in scientific research to assure the veracity and integrity of research data, especially in areas like satellite imaging and medical imaging. To ensure the authenticity and dependability of scientific findings, picture forgery detection algorithms are employed to spot any tampering of research photos. As digital images are increasingly used in a variety of applications, the breadth of image forgery detection is also expanding. To maintain the validity and integrity of digital images in numerous industries, image forgery detection algorithms are continuously improving to keep up with the development of new image modification techniques.

Several techniques and algorithms that can correctly and effectively identify modified photos are used to detect image counterfeiting. The solution is multifaceted and includes various processes, such as feature extraction from images, classification, and image preparation. Image preprocessing, which entails improving the image's quality and eliminating any artefacts or noise that can hinder the detection of alterations, is the initial step in the detection of image forgeries. Image denoising, filtering, and color correction are examples of preprocessing techniques. Identifying and extracting pertinent information from an image that can be utilized to differentiate between modified and genuine images is the second step in the detection of image forgeries. Examples of feature extraction techniques include local binary patterns (LBP), scale-invariant feature transforms (SIFT), and histograms of oriented gradients (HOG). The classification phase in image forgery detection involves using a machine learning or deep learning algorithm to analyze the extracted features and determine whether an image is authentic or modified. Support vector machines (SVMs), decision trees, and CNNs are examples of classification techniques.

Detecting specific forms of image manipulation, such as deep fakes, copy-move alterations, and steganography, demands the creation of specialized algorithms. Developing these techniques is essential, as forgers often employ advanced and complex strategies to conceal their modifications. In general, the development of reliable and effective methods and algorithms that can precisely identify a variety of image alterations is required to solve the problem of image fraud detection, ensuring the validity and integrity of digital images in diverse applications. Figure 5.1 shows the images with different forgeries.

To spot modifications in an image, traditional image processing methods including median filtering, wavelet transformation, and edge detection are applied. These methods are based on spotting artefacts that the forger introduced during the alteration procedure. Although they move quite quickly, they are not very accurate, and they are frequently targeting clever forgeries. Approaches based on certain elements of a picture, such as its texture, color, and edge information, are extracted and analyzed in

Figure 5.1 Different types of images with different forgeries.

feature-based approaches. Then, these characteristics are applied to identify image modifications. Although feature-based methods are often accurate, they could not be effective against complex manipulations like deepfakes. In conclusion, standard image processing methods are quick but only partially accurate. Although accurate, feature-based techniques might not perform well for complex modifications. Machine learning techniques offer precision and efficiency for detecting complex manipulations, while deep learning methods achieve high accuracy but require substantial processing power and extensive training data. The choice of technique depends on the application's requirements and the complexity of the image modifications to be detected. A workflow of image forgery detection using CNN is shown in Figure 5.2.

5.2 LITERATURE SURVEY

Image forgery detection is an essential area of research that focuses on identifying manipulations in digital images. With the rise of social media and digital communication, the authenticity of images has become increasingly questionable, leading to the need for reliable detection methods. Image forgery can take many forms, including splicing, copy-move, and image retouching, all of which aim to mislead viewers by altering visual information. In a world where misinformation can spread rapidly, the ability to detect altered images is crucial. For instance, news organizations rely on authentic images to convey accurate information, and forensic experts use image analysis in legal contexts to validate evidence. The advent of

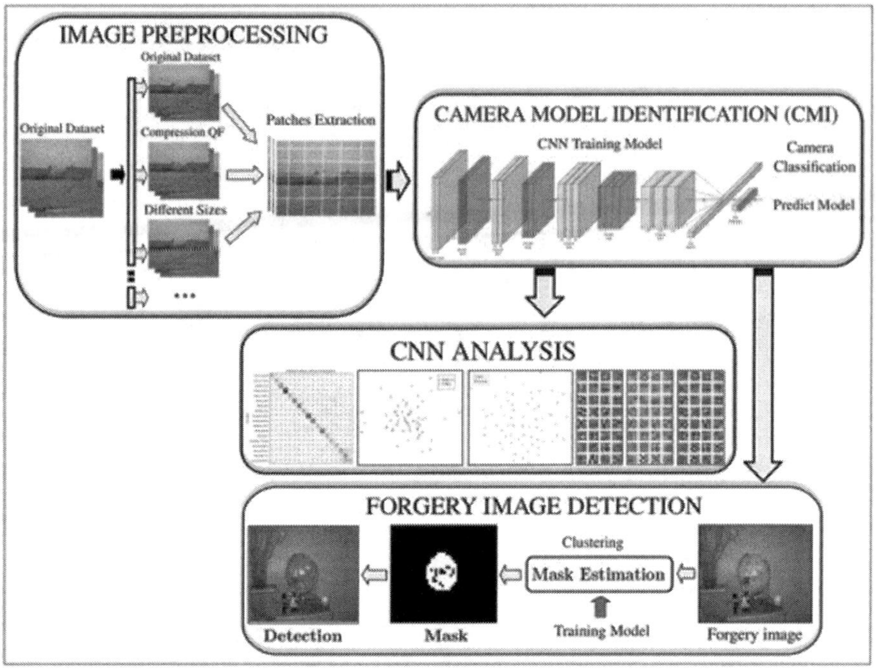

Figure 5.2 Working of image forgery detection using CNN.

deep learning techniques, especially CNNs, has transformed image forgery detection, significantly enhancing the accuracy and efficiency in identifying tampered images [9].

Research work done by Fan et al. [10] shows several classifiers used in image forgery detection, each offering unique advantages and limitations. The main classifiers applied in this research work are as follows:

- Convolutional neural networks (CNNs) in image forgery detection: CNNs have become the benchmark in image processing tasks because their architecture is uniquely suited to capturing spatial hierarchies within images. CNNs automatically learn features from raw pixel values, enabling them to capture intricate details and patterns that may indicate forgery. Studies have demonstrated that CNNs outperform traditional classifiers in various image-related tasks, including forgery detection.
- Support vector machines (SVMs): SVMs are a popular choice for classification tasks due to their effectiveness in high-dimensional spaces. However, SVMs require careful feature engineering and are

often limited in their ability to handle complex visual data, which can reduce their effectiveness in forgery detection.

- Random forests in image forgery detection: Random forests, an ensemble learning technique, combine multiple decision trees to enhance classification accuracy. Though they are resistant to overfitting and can manage diverse datasets, they face challenges with the high dimensionality of image data, which can result in less effective performance in forgery detection tasks.
- k-Nearest neighbors (k-NN): k-NN is a straightforward classification method that relies on distance metrics to classify instances based on their neighbors. However, k-NN is computationally expensive for large datasets and often fails to capture complex relationships in high-dimensional image data, making it less suitable for forgery detection.
- Ensemble methods: These methods combine multiple classifiers to leverage their strengths. However, they can introduce additional complexity and computational overhead, making them less appealing compared to the streamlined approach of using CNNs.

Although several classifiers have been developed and evaluated for this task and out of all the classifiers, CNNs stand out for their exceptional ability to learn intricate features directly from image data. As research in this field progresses, the incorporation of advanced machine learning techniques is expected to further improve the reliability and accuracy of image forgery detection methods.

In 2023, CNNs [11] in image forgery detection have emerged as the foundation of image processing tasks, especially in image forgery detection. A specialized architecture is built to recognize spatial hierarchies within images, enabling them to effectively detect subtle alterations that may signal forgery. The detail of work is as follows:

- Architecture and Feature Learning: CNNs consist of layers including convolutional filters, pooling layers, and fully connected layers. The convolutional layers play a crucial role, applying a range of filters to input images to detect edges, textures, and patterns across various scales. This multilayered approach allows CNNs to learn complex features without manual feature extraction, a significant advancement over traditional classifiers like SVMs or random forests. As highlighted in recent studies, the ability of CNNs to adaptively learn features makes them particularly effective in scenarios where the manipulations in an image are subtle or where the forgery is crafted to appear as realistic as possible. This adaptability results in enhanced detection capabilities, most importantly while dealing with huge datasets.

- Training and Generalization: The training process for CNNs generally involves using a large, labeled dataset to adjust the network's weights through backpropagation. This allows the model to minimize a loss function that measures the difference between predicted outputs and true labels. According to a study by Zhang et al., well-optimized CNNs can achieve accuracy rates greater than 95% in detecting different types of image forgery. Moreover, the generalization capability of CNNs allows them to perform effectively even on unseen data. This is particularly important in forgery detection, where the variety of potential manipulations can be vast. Techniques such as data augmentation (e.g., rotating, scaling, and flipping images) are often employed during training to enhance this generalization, making CNNs robust against overfitting.
- Comparative Performance with Other Classifiers: Research has consistently shown that CNNs outperform traditional machine learning classifiers in forgery detection tasks. For example, Kandasamy and Sridhar (2023) reported that while SVMs and random forests can achieve reasonable accuracy, they often fail to detect nuanced forgeries, leading to higher false-negative rates. This is largely because these classifiers rely heavily on predefined features and lack the capacity to learn from raw data.

Furthermore, ensemble methods, while combining the strengths of multiple classifiers, often add complexity and computational overhead without significantly improving accuracy compared to a well-trained CNN. The streamlined architecture of CNNs allows for efficient processing of high-dimensional image data, further solidifying their position as the leading choice for forgery detection.

Recent advancements in deep learning have significantly improved image forgery detection, particularly using CNNs. Baccari [9] demonstrated that deep CNN architectures can effectively identify various types of image tampering by learning hierarchical representations of manipulated features. This approach eliminates the need for manual feature engineering, making CNNs a powerful alternative to traditional classifiers.

- Real-Time Applications: Kim [12] proposed augmented CNNs optimized for real-time image forgery detection. By leveraging hardware acceleration techniques such as GPUs, CNN-based models can be deployed in practical scenarios requiring instant forgery identification, benefiting fields like journalism, law enforcement, and digital forensics.
- Hybrid Approaches: Smith and Anwar [11] explored the integration of CNNs with generative adversarial networks (GANs) to enhance forgery detection. This method not only improves detection accuracy

but also enables the generation of synthetic images to refine detection algorithms.

- Feature Learning vs. Traditional Methods: Weng et al. [13] compared various deep learning architectures with conventional classifiers, concluding that traditional models such as SVM and k-nearest neighbors (k-NN) struggle with complex image manipulations due to their dependence on handcrafted features.
- Challenges of High Dimensionality: Nguyen and Ma [14] highlighted that SVM and Random Forest classifiers perform poorly when handling high-dimensional image data, often suffering from the curse of dimensionality. CNNs, in contrast, efficiently process large-scale images using convolutional layers that capture spatial dependencies.
- Computational Complexity: Jiang et al. [15] examined the computational demands of different classification models and noted that while SVMs require intensive computations, CNNs achieve superior performance by leveraging optimized architectures and parallel computing.
- Robustness Against Noise: Wilson [16] proposed contrastive learning techniques to enhance CNN-based forgery detection against noisy inputs. Unlike k-NN, which is highly sensitive to variations in pixel intensity, CNNs learn robust feature representations that improve classification accuracy even in the presence of noise.
- Spatial Relationships in Forgery Detection: Edwards [17] demonstrated that CNNs surpass traditional classifiers by effectively capturing spatial correlations between pixels. This is crucial for detecting subtle alterations, such as splicing or copy-move forgeries, which often go undetected by SVMs and decision tree-based models.

CNNs have revolutionized image forgery detection by offering superior accuracy, automation, and robustness. As research in deep learning continues, the integration of CNNs with hybrid models and emerging techniques is expected to further enhance the reliability of forgery detection systems, strengthening the integrity of digital media.

Several classifiers have been employed in image forgery detection, with k-nearest neighbors (k-NN) and ensemble methods such as random forests and gradient boosting being prominent alternatives to CNNs. However, these methods exhibit certain limitations that affect their efficiency in detecting forged images. Researchers like Jiang et al. [15] have analysed the performance of different classifiers, highlighting challenges in k-NN and ensemble approaches when compared to CNN-based models.
Limitations of k-NN classifier:

- Sensitivity to Feature Scaling and Noise – As noted by Yasir and Chang [18], k-NN's reliance on distance metrics makes it highly sensitive to feature scaling and noise. In image forgery detection, varying pixel intensity values can lead to misclassification if features are not properly normalized.

- High Computational Cost – Weng et al. [13] demonstrated that k-NN requires storing the entire training dataset and computing distances for each query, making it computationally expensive, especially for large-scale image datasets. This drawback limits its applicability in real-time forensic tasks.
- Inefficiency in High-Dimensional Spaces – As dimensionality increases, the effectiveness of k-NN diminishes due to the curse of dimensionality (Gao and Munir [19]). In image analysis, where feature space is large, k-NN often suffers from poor classification accuracy.

Limitations of Ensemble Classifiers:

- Complexity and Interpretability – Wilson [16] pointed out that ensemble methods like Random Forests aggregate multiple models, making it difficult to interpret individual predictions. In forensic investigations, transparency is crucial, which CNNs address more effectively.
- Risk of Overfitting – Nguyen and Ma [14] found that ensemble classifiers, despite their robustness, are prone to overfitting when models become too complex, leading to degraded performance on unseen data. CNNs, with their hierarchical learning structure, generalize better.
- Computational Demand – Hassan et al. [20] highlighted that ensemble methods require extensive computational resources, particularly in training, making them less suitable for real-time forgery detection compared to CNNs.

Comparisons with CNNs:

- Feature Learning vs. Manual Feature Engineering – Baccari [9] demonstrated that CNNs automatically extract hierarchical features from raw images, eliminating the need for extensive manual feature engineering required in k-NN and ensemble classifiers.
- Robustness to Noise and Variability – Lu and Chin [21] found that CNNs are more resilient to noise due to their use of pooling layers, which help generalize features across different image conditions, whereas traditional classifiers struggle with variations in input data.
- Handling of High-Dimensional Data – Edwards [17] showed that CNNs effectively reduce dimensionality through learned convolutions, preserving important image details while discarding irrelevant

information. Traditional classifiers, on the other hand, struggle with increasing dimensions, leading to performance degradation.

While classifiers like k-NN and ensemble methods have been widely applied to image forgery detection, their limitations—such as sensitivity to noise, computational inefficiency, and complexity—reduce their effectiveness. CNNs, leveraging deep learning for automatic feature extraction, offer a more robust and scalable approach, outperforming traditional classifiers in identifying forged images.

CNNs have become the dominant approach for image forgery detection due to their ability to learn intricate patterns directly from raw image data. Unlike traditional methods that rely on manual feature extraction, CNNs leverage hierarchical layers to automatically detect forgeries, making them highly adaptable across different datasets [15].

- CNNs excel in automatic feature extraction, as demonstrated by Baccari [9], who showed that deep convolutional networks can identify fine-grained tampering patterns without manual intervention. This eliminates the limitations of handcrafted features, improving detection accuracy.
- Robustness to variability and noise is another advantage. Fan et al. [10] found that CNN-based models outperform traditional classifiers like SVMs and k-NN when dealing with compression artifacts, lighting variations, and distortions, ensuring high reliability in real-world applications.
- Scalability and generalization make CNNs effective in handling large datasets. Smith and Anwar [11] highlighted how GPU-accelerated CNN models maintain strong performance on unseen data, reducing overfitting risks and improving model reliability.
- CNNs are also adaptable to different forgery types. Zhong [22] demonstrated that hierarchical CNN models could be fine-tuned to detect copy-move, splicing, and image synthesis forgeries, whereas traditional methods require extensive manual adjustments to adapt to new forgery techniques.
- The integration of CNNs with advanced techniques enhances detection performance. Nguyen and Ma [14] explored how transfer learning and ensemble CNN approaches significantly boost accuracy, allowing pre-trained models to perform well even with limited datasets.

These capabilities position CNNs as the leading choice for image forgery detection, providing superior accuracy, adaptability, and robustness in forensic applications [15].

5.3 METHODOLOGY

This section discusses how image forgery detection operates in digital photos. Image preprocessing, feature extraction, and classification are some of the processes in the procedure which are required, as shown in Figure 5.3. The first step is to prepare the image in the initial stage in the detection of image forgeries. This procedure comprises enhancing the image's quality and eliminating any noise or artefacts that can make it more difficult to spot changes. Techniques used for preparing images include filtering, color correction, and picture denoising.

The second step in image forgery detection involves identifying and extracting relevant features from the image that can help differentiate between authentic and altered images. Techniques such as LBPs, SIFTs, and HOGs are commonly used for this purpose.

Classification is the final step in detecting image forgery, where the goal is to determine if an image is authentic or manipulated. A classifier is trained on a dataset of genuine and altered images to learn the distinguishing features. Machine learning techniques, such as SVMs, decision trees, and neural networks, can then be used to classify the images based on these learned characteristics.

There are numerous techniques for identifying sorts of picture manipulation, such as copy-move manipulation, splicing, and deep fakes, in addition to these processes. These techniques might involve inspecting the image for indications of manipulation or distortion or analyzing it for irregularities in the lighting, shadows, and edges.

To spot picture fraud, it is usually necessary to examine the image's content for any discrepancies or anomalies that might point to tampering. A logical database for picture fraud detection would have to meet the following specifications to accomplish this effectively:

- Image metadata: The database should contain information about the image, such as the file type, the equipment used to take it, and the date it was created. If the image has been edited or manipulated, this information may help identify it.
- Image features: Such as color histograms, edge information, and texture details, should be stored in a database. This information can be used to detect irregularities, like shifts in texture or color, which may indicate potential tampering.
- Image comparison: Algorithms for comparing photos, such as pixel-by-pixel, feature-based, and machine learning-based comparison, should be included in the database. These algorithms can be applied to compare photographs, which can determine whether an image is tempered or not.

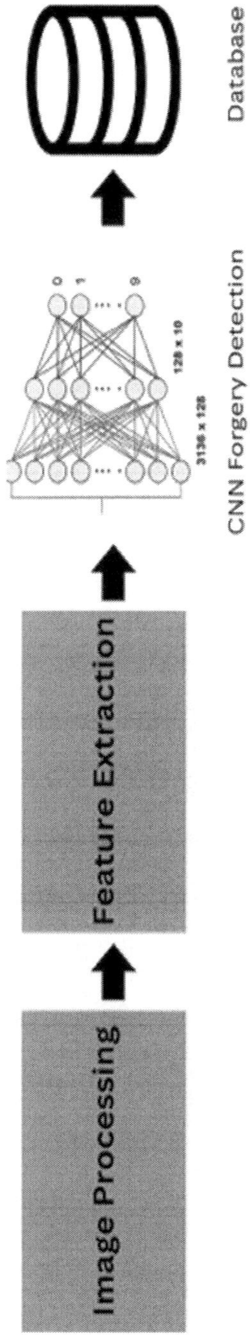

Figure 5.3 Architecture for image forgery detection.

- Reference database: The database needs to have a reference database of verified images that may be used as a standard against which the image in question can be measured. This can assist in identifying any differences or inconsistencies that can lead to forgery.
- Database should have an audit trail that records all activity relating to the image, including when it was taken, who had access to it, and any modifications that were made. This makes it easier to spot any shady activity connected to the photograph.

A logical database for picture forgery detection should, in general, have a variety of features and algorithms that can be used to compare and analyze photos to find any discrepancies or anomalies that would suggest manipulation. CNNs are widely used for image forgery detection because of their ability to identify complex features and patterns within images. Below is shown a use case diagram of a general approach for building a CNN-powered image forgery detection system (Figure 5.4).

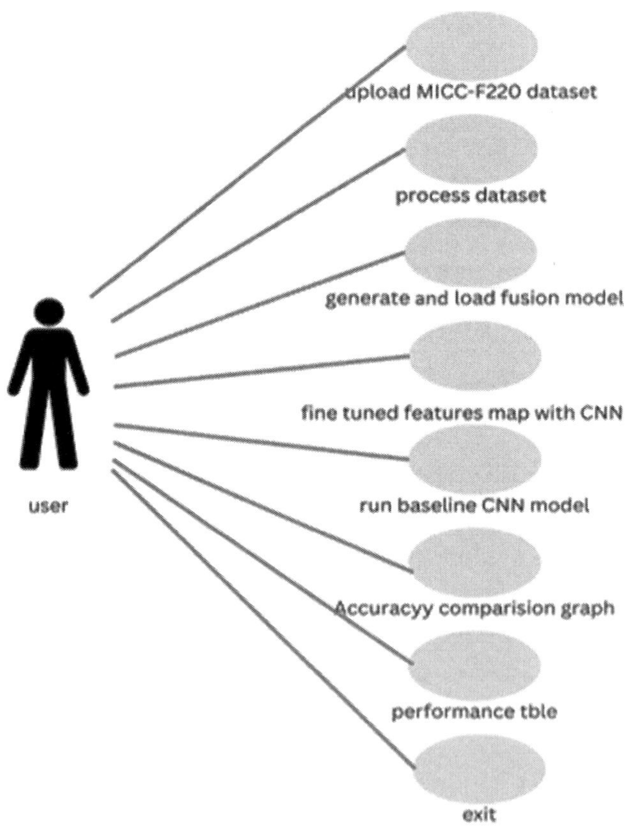

Figure 5.4 Use case of the CNN-based image fraud detection system.

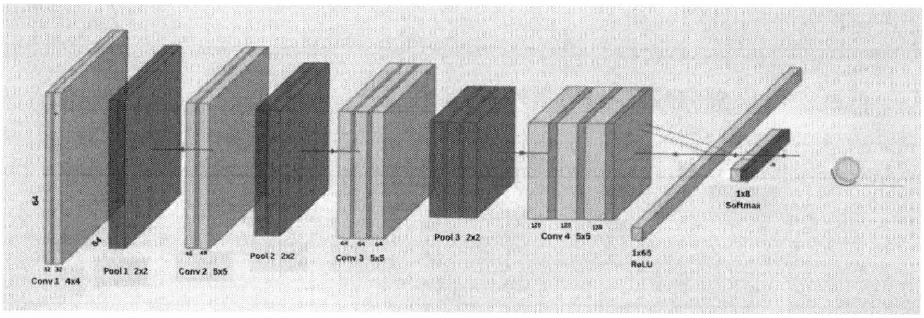

Figure 5.5 Different layers in CNN model.

Algorithm for CNN-based Image Fraud Detection System

- Data collection and preprocessing: Collect a large and varied dataset containing both authentic and manipulated images. Each image should be labeled to indicate whether it is genuine or altered. It's also important to divide the dataset into training, validation, and test sets to ensure proper model evaluation.
- Designing a network architecture Select the best CNN architecture for the picture forgery detection challenge, such as VGG, ResNet, or Inception. Change the architecture as needed to make it more effective at the task (See Figure 5.5).
- Using a suitable optimization approach, such as stochastic gradient descent (SGD), train CNN using the labelled dataset. CNN must have the ability to tell the difference between real and altered visuals.
- To enhance the CNN's performance on the validation set, try adjusting different hyperparameters such as the learning rate, batch size, and regularization strength through experimentation.
- Testing the CNN: Check the CNN's performance on the test set to see how well it can spot fake images. On the testing set, the CNN should achieve high levels of accuracy, precision, and recall.
- Installation: Install the system for spotting fake images in a real-world setting. New photos should be processed by the system, and any instances of forgery should be correctly identified.
- Continuous improvement: To keep the picture forgery detection system's accuracy and dependability, continuously assess its performance and make appropriate adjustments.

In general, picture forgery detection functions by examining digital photos using a variety of approaches and algorithms to find any indications of manipulation or change, hence maintaining the authenticity and integrity of digital images in a variety of applications.

5.4 IMPLEMENTATION AND RESULT

The implementation of an image forgery detection system using CNNs involves several key steps, including the following:

- Image Preprocessing: The images must first be preprocessed by resizing them to a standard dimension, converting them into the appropriate format (e.g., RGB or grayscale), and normalizing the pixel values to a consistent scale.
- Building the CNN Model: The model should be designed to accept input images that match the preprocessed size. The architecture typically consists of multiple convolutional layers, followed by pooling layers, and finally, fully connected layers. The model must be optimized to achieve high accuracy in detecting image forgeries.
- Training the CNN Model: During the training process, preprocessed images are fed into the CNN model. The model's weights are updated through backpropagation to minimize the loss function and improve performance.
- Testing the CNN Model: The trained model is evaluated on a separate labeled dataset that was not used during training. Metrics like accuracy, precision, recall, and F1 score are calculated to assess the model's effectiveness in detecting forgeries.
- Hyperparameter Tuning: To improve the CNN's performance, key hyperparameters—such as the learning rate, batch size, and number of epochs—should be fine-tuned. Techniques like grid search or random search can be employed to optimize these hyperparameters.
- The trained CNN model can be used to detect fake images in real time in a production environment. The model's output can be used after preprocessing the input photos using the same methods as during training.
- System monitoring and development: The effectiveness of the picture forgery detection system should be regularly assessed and enhanced. This can be achieved by gathering user feedback, examining how the model performs when applied to fresh data, and modifying the model as required.
- Overall, developing and training the CNN model, testing and refining the model, deploying the system, and continuously monitoring and optimizing its performance are all steps in the implementation of an image fraud detection system utilizing CNNs.

5.4.1 Image forgery detection using CNN python (trained using CASIA2 dataset)

- Begin by importing all necessary libraries and mounting Google Drive if you're using Google Colab (Figure 5.6a).
- Convert an image to an error level analysis (ELA) image (Figure 5.6b).
- Prepare an image by converting it to ELA and resizing it (Figure 5.6c).
- Dataset Preparation: Load and preprocess the dataset. Randomly select and shuffle the images for both genuine and forged categories (Figure 5.7a and b).
- Train-Test Split process: Divide the dataset into separate training and validation subsets (Figure 5.8).
- Define and compile the CNN model (Figure 5.9).
- Train the CNN model with early stopping and Plot Training History (Figure 5.10a and b).
- Evaluate the model using a confusion matrix and performance metrics (Figure 5.11a). Predict the class of new images and calculate the accuracy on a test set of fake and real images (Figure 5.11b).

The implementation of an image forgery detection system employing CNNs may encounter several problems. Here are some important points to think about:

- Dataset bias: Bias in the training dataset may have an impact on how well the CNN model performs. For instance, the model may become biased towards one class or the other if the training dataset contains

```
import numpy as np
import matplotlib.pyplot as plt
from sklearn.model_selection import train_test_split
from sklearn.metrics import confusion_matrix
from tensorflow.keras.utils import to_categorical
from keras.models import Sequential
from keras.layers import Dense, Flatten, Conv2D, MaxPool2D, Dropout
from keras.optimizers import Adam
from keras.preprocessing.image import ImageDataGenerator
from keras.callbacks import EarlyStopping
from PIL import Image, ImageChops, ImageEnhance
import os
import itertools
from google.colab import drive

# Mount Google Drive
drive.mount('/content/gdrive')
```

(a)

```
def convert_to_ela_image(path, quality):
  temp_filename = 'temp_file_name.jpg'
  ela_filename = 'temp_ela.png'

  image = Image.open(path).convert('RGB')
  image.save(temp_filename, 'JPEG', quality=quality)
  temp_image = Image.open(temp_filename)

  ela_image = ImageChops.difference(image, temp_image)

  extrema = ela_image.getextrema()
  max_diff = max([ex[1] for ex in extrema])
  if max_diff == 0:
    max_diff = 1
  scale = 255.0 / max_diff

  ela_image = ImageEnhance.Brightness(ela_image).enhance(scale)

  return ela_image
```

(b)

```
image_size = (128, 128)

def prepare_image(image_path):
    return np.array(convert_to_ela_image(image_path, 90).resize(image_size)).flatten() / 255.0
```

(c)

Figure 5.6 Import libraries and prepare images.

```
# Initialize lists to store images and labels
X = [] # ELA converted images
Y = [] # 0 for fake, 1 for real

# Set paths to genuine and forged images
genuine_path = '/kaggle/input/casia-dataset/CASIA2/Au/'
forged_path = '/kaggle/input/casia-dataset/CASIA2/Tp/'

# Load genuine images
for dirname, _, filenames in os.walk(genuine_path):
    for filename in filenames:
        if filename.endswith('jpg') or filename.endswith('png'):
            full_path = os.path.join(dirname, filename)
            X.append(prepare_image(full_path))
            Y.append(1)
            if len(Y) % 500 == 0:
                print(f'Processing {len(Y)} images')

# Shuffle and reduce dataset to 2100 genuine images
random.shuffle(X)
X = X[:2100]
Y = Y[:2100]
```

(a)

```
# Load forged images
for dirname, _, filenames in os.walk(forged_path):
    for filename in filenames:
        if filename.endswith('jpg') or filename.endswith('png'):
            full_path = os.path.join(dirname, filename)
            X.append(prepare_image(full_path))
            Y.append(0)
            if len(Y) % 500 == 0:
                print(f'Processing {len(Y)} images')

# Convert lists to NumPy arrays and reshape
X = np.array(X)
Y = to_categorical(Y, 2)
X = X.reshape(-1, 128, 128, 3)
```

(b)

Figure 5.7 Data preparation and pre-processing.

```
X_train, X_val, Y_train, Y_val = train_test_split(X, Y, test_size=0.2, random_state=5)
print(len(X_train), len(Y_train))
print(len(X_val), len(Y_val))
```

Figure 5.8 Train-test data split.

an excessive amount of legitimate or altered photos. A diverse array of photos must be gathered, and the classes must be as evenly distributed as feasible, to overcome this problem.

- Limited data availability: For rare instances of picture forgeries, it can be difficult to compile a sizable collection of labelled images. The CNN model may overfit to the training dataset if the dataset that is trained with is too small, and it may perform badly on fresh data. This problem can be solved by using data augmentation techniques to fictitiously expand the dataset and strengthen the model's robustness.
- Computing power: Using a sizable dataset to train a CNN model might demand a lot of processing power, such as powerful GPUs and lots of RAM. This can be costly and might call for specialized gear. Cloud computing platforms can be utilized to rent compute resources on an as-needed basis to solve this problem.
- Interpretability: It can be difficult to grasp how CNN models generate their predictions because of their difficulty in interpretation. This might be a problem in situations where explainability is crucial. To visualize which areas of the image are most crucial for the model's

```
def build_model():
    model = Sequential()
    model.add(Conv2D(filters=32, kernel_size=(5, 5), padding='valid', activation='relu', input_shape=(128, 128, 3)))
    model.add(Conv2D(filters=32, kernel_size=(5, 5), padding='valid', activation='relu'))
    model.add(MaxPool2D(pool_size=(2, 2)))
    model.add(Dropout(0.25))
    model.add(Flatten())
    model.add(Dense(256, activation='relu'))
    model.add(Dropout(0.5))
    model.add(Dense(2, activation='softmax'))
    return model

model = build_model()
model.summary()

# Compile the model
epochs = 30
batch_size = 32
init_lr = 1e-4
optimizer = Adam(lr=init_lr, decay=init_lr/epochs)

model.compile(optimizer=optimizer, loss='binary_crossentropy', metrics=['accuracy'])
```

Figure 5.9 CNN model.

```
early_stopping = EarlyStopping(monitor='val_accuracy',
                               min_delta=0,
                               patience=2,
                               verbose=0,
                               mode='auto')

hist = model.fit(X_train,
                 Y_train,
                 batch_size=batch_size,
                 epochs=epochs,
                 validation_data=(X_val, Y_val),
                 callbacks=[early_stopping])

# Save the trained model
model.save('model_casia_run1.h5')
```

(a)

```
# Plot the loss and accuracy curves for training and validation
fig, ax = plt.subplots(2, 1, figsize=(10, 10))

# Plot training & validation loss values
ax[0].plot(hist.history['loss'], color='b', label="Training loss")
ax[0].plot(hist.history['val_loss'], color='r', label="Validation loss")
ax[0].set_title('Model Loss')
ax[0].set_ylabel('Loss')
ax[0].set_xlabel('Epoch')
ax[0].legend(loc='upper left')

# Plot training & validation accuracy values
ax[1].plot(hist.history['accuracy'], color='b', label="Training accuracy")
ax[1].plot(hist.history['val_accuracy'], color='r', label="Validation accuracy")
ax[1].set_title('Model Accuracy')
ax[1].set_ylabel('Accuracy')
ax[1].set_xlabel('Epoch')
ax[1].legend(loc='upper left')

plt.show()
```

(b)

Figure 5.10 CNN early stopping and plotting.

predictions, techniques like saliency maps and gradient-based attribution can be employed to solve this problem.

- Assaults from the opposite direction: CNN models are susceptible to assaults from the opposite direction, in which a minor perturbation is injected into the input image to make the model predict incorrectly.

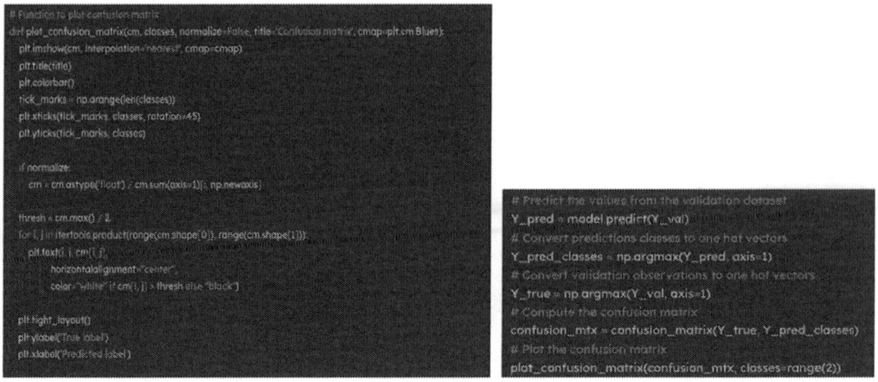

```
# Function to plot confusion matrix
def plot_confusion_matrix(cm, classes, normalize=False, title='Confusion matrix', cmap=plt.cm.Blues):
    plt.imshow(cm, interpolation='nearest', cmap=cmap)
    plt.title(title)
    plt.colorbar()
    tick_marks = np.arange(len(classes))
    plt.xticks(tick_marks, classes, rotation=45)
    plt.yticks(tick_marks, classes)

    if normalize:
        cm = cm.astype('float') / cm.sum(axis=1)[:, np.newaxis]

    thresh = cm.max() / 2.
    for i, j in itertools.product(range(cm.shape[0]), range(cm.shape[1])):
        plt.text(j, i, cm[i, j],
                 horizontalalignment="center",
                 color="white" if cm[i, j] > thresh else "black")

    plt.tight_layout()
    plt.ylabel('True label')
    plt.xlabel('Predicted label')
```

(a)

```
# Predict the values from the validation dataset
Y_pred = model.predict(X_val)
# Convert predictions classes to one hot vectors
Y_pred_classes = np.argmax(Y_pred, axis=1)
# Convert validation observations to one hot vectors
Y_true = np.argmax(Y_val, axis=1)
# Compute the confusion matrix
confusion_mtx = confusion_matrix(Y_true, Y_pred_classes)
# Plot the confusion matrix
plot_confusion_matrix(confusion_mtx, classes=range(2))
```

(b)

Figure 5.11 Results code.

In applications where security is crucial, this can be a problem. This problem can be solved by using methods like adversarial training and input preprocessing to strengthen the model's defense against hostile attacks.

5.5 RESULT ANALYSIS

The test dataset can be divided into multiple groups depending on the type of counterfeit (e.g., copy-move, splicing, or retouching) to know the system's capacity to find out various forgeries. This allows for a distinct assessment of the system's performance for each category. A confusion matrix shown in Figure 5.12 illustrates the performance of a classification model on a test dataset with known true values. It provides a visual representation of the algorithm's effectiveness.

Accuracy: The overall accuracy of the model, which is the ratio of correctly predicted instances (both true positives and true negatives) to the total instances is calculated. The accuracy of 87.0% indicates that the model is generally performing well.

$$\text{Accuracy} = \frac{\text{TP} + \text{TN}}{\text{TP} + \text{TN} + \text{FP} + \text{FN}} = \frac{370 + 357}{370 + 357 + 46 + 62}$$

$$= \frac{727}{835} \approx 0.870(\text{or } 87.0\%)$$

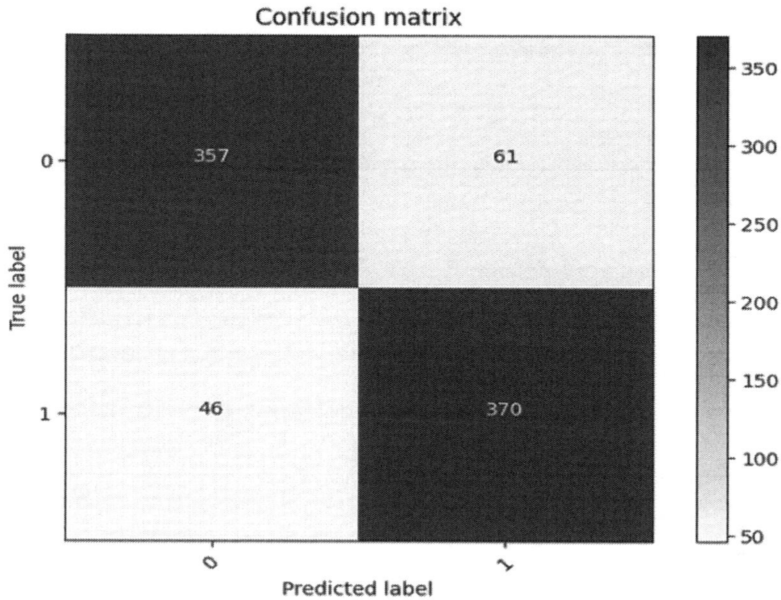

Figure 5.12 Confusion matrix.

Precision (for Real): The ratio of correctly predicted real images to the total predicted real images. The precision of 88.9% indicates that when the model predicts an image as real, it is correct 88.9% of the time.

$$\text{Precision} = \frac{\text{TP}}{\text{TP} + \text{FP}} = \frac{370}{370 + 46} \approx 0.889(\text{or } 88.9\%)$$

Recall (Sensitivity or True Positive Rate, for Real): The ratio of correctly predicted real images to the total actual real images. The recall of 85.6% indicates that the model correctly identifies 85.6% of the real images.

$$\text{Recall} = \frac{\text{TP}}{\text{TP} + \text{FP}} = \frac{370}{370 + 62} \approx 0.856(\text{or } 85.6\%)$$

F1 Score (for Real): The harmonic mean of precision and recall. F1 Score of 87.2% strikes a balance between precision and recall, offering a unified metric that accounts for both false positives and false negatives.

$$F1 \cdot \text{Score} = 2 \times \frac{\text{Precision} \times \text{Recall}}{\text{Precision} + \text{Recall}}$$

$$= 2 \times \frac{0.889 \times 0.856}{0.889 + 0.856} \approx 0.872 (\text{or } 87.2\%)$$

Specificity: The ratio of correctly predicted fake images to the total actual fake images. The specificity of 88.6% indicates that the model correctly identifies 88.6% of the fake images.

$$\text{Specificity} = \frac{\text{TN}}{\text{TN} + \text{FP}} = \frac{357}{357 + 46} \approx 0.886 (\text{or } 88.6\%)$$

False Positive Rate (FPR): The ratio of incorrectly predicted real images to the total actual fake images. The false positive rate of 11.4% shows the proportion of fake images incorrectly classified as real.

$$\text{FPR} = \frac{\text{FP}}{\text{FP} + \text{TN}} = \frac{46}{46 + 357} \approx 0.114 (\text{or } 11.4\%)$$

The loss and accuracy curves for training and validation loss: The training loss and validation loss graphs provide critical insights into the performance and generalization ability of your CNN model for forgery image detection.

As shown in Figure 5.13, the training loss starts at 0.6 which indicates the high error on the training data. This is expected as the model starts learning from scratch. The training loss decreases steadily over epochs while reaching 0.43. This indicates that the model is learning from the training data and improving its performance. Validation loss starts at 0.49 and decreases to 0.45, indicating that the model is initially learning and generalizing well to unseen data. The validation loss continues to decrease to 0.44, suggesting further improvement in the model's generalization. However, the slight increase and subsequent return to 0.45 may indicate that the model is struggling to maintain its performance or is overfitting to the validation data. The training accuracy starts at 0.6 and increases significantly to 0.85. This indicates that the model is learning from the training data and improving its ability to correctly classify samples. After reaching 0.85, the training accuracy stabilizes around this value, fluctuating slightly between 0.84 and 0.85. This suggests that the model has learned most of the patterns present in the training data and is performing consistently. The validation accuracy starts relatively high at 0.9, indicating that the model initially generalizes well to unseen data. The validation accuracy decreases to 0.86 and then to

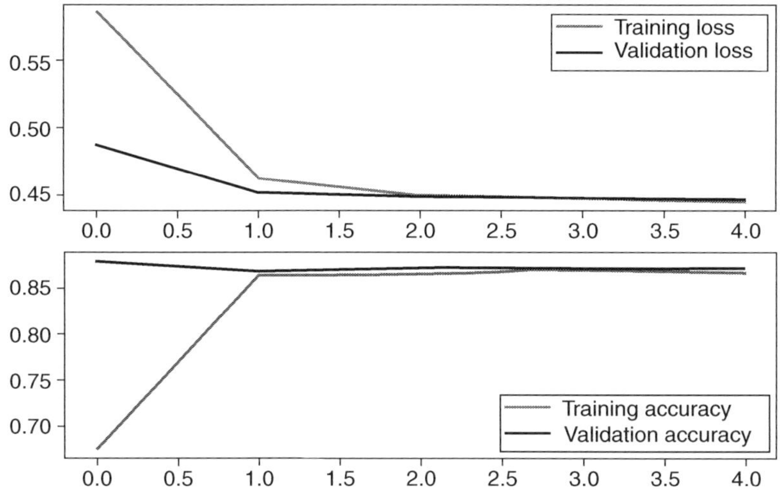

Figure 5.13 Accuracy and loss graph.

0.84 before returning to 0.86. This suggests that the model may be struggling to generalize to unseen data, possibly due to overfitting or the presence of outliers or noisy data points in the validation set.

5.6 FUTURE WORK

This research work presents a reliable and comprehensible CNN-based model for image forgery detection. Despite of great accuracy, CNN-based models can be challenging to comprehend and subject to hostile attacks. One of the primary limitations lies in the black-box nature of deep learning models, making it difficult to understand the decision-making process behind forgery detection. Future efforts should focus on improving the explainability and interpretability of CNN models through visualization techniques, attention mechanisms, and interpretable architectures.

Additionally, robustness against adversarial attacks remains a crucial area for improvement. Adversarial manipulations can deceive CNN models by introducing imperceptible perturbations, leading to misclassification. Enhancing model resilience through adversarial training, defensive distillation, and anomaly detection techniques can significantly improve reliability in forensic applications.

Furthermore, cross-domain generalization is essential for deploying CNN-based forgery detection models in real-world scenarios. Models trained on specific datasets often struggle to generalize across different forgery techniques and image sources. Leveraging domain adaptation, transfer learning, and meta-learning can help CNNs adapt to diverse datasets and maintain high detection accuracy.

Another promising direction involves integrating CNNs with hybrid approaches, combining deep learning with traditional forensic techniques. Fusion models—leveraging both deep feature extraction and handcrafted forensic features—can improve robustness and interpretability.

Lastly, the ethical and legal implications of image forgery detection must be considered. As forensic AI models become more advanced, ensuring bias-free decision-making, transparency, and compliance with legal standards will be critical for their adoption in law enforcement and media forensics.

Future research can concentrate on creating more resilient models that are more interpretable and less vulnerable to attacks.

- Increasing the diversity and quality of training data: The training data's quality and diversity have a significant impact on the accuracy of CNN-based models. To enhance the effectiveness of these models, future research can concentrate on creating datasets that are larger and more varied.
- Integration of numerous modalities: In addition to looking at photos, other types of modalities, such audio and text, can also be used to detect image forgeries. Future research can concentrate on merging many modalities to create more thorough counterfeit detection systems that can examine various media types.
- CNN-based models can be computationally expensive and may not be appropriate for deployment on mobile or low-power devices, making them unsuitable for mobile and low-power device modelling. Future research can concentrate on creating models that can operate on these gadgets without sacrificing accuracy.
- Creating CNN-based models to detect video forgery is more difficult than detecting image forgery, despite the fact that CNN-based models have been successful at doing so. Future research can concentrate on creating CNN-based algorithms that can recognize video piracy, including deepfake videos.
- Creating models for real-time forgery detection: Real-time image fraud detection is essential for many applications, including social networking platforms and video streaming. Future research can concentrate on creating CNN-based models that can be incorporated into these systems and can instantly detect forgeries.
- Making multimodal forgery detection models multimodal forgery detection entails spotting fakes that were produced by fusing various media, such audio and visual. Future research can concentrate on creating CNN-based algorithms that can recognize multi-modal forgeries and offer more thorough analysis.

REFERENCES

[1] H. S. Yoon and G. H. Park, "Modeling Image Forgery Localization using CNN and Transfer Learning," *IEEE Transactions on Artificial Intelligence*, vol. 5, no. 3, pp. 159–172, 2023.

[2] J. A. Baek, "Detecting GAN-Generated Fake Images: A CNN Approach," In *Proc. IEEE Int. Conf. Image Processing*, 2023, pp. 1027–1036.

[3] Choudhary, A., & Arora, A. (2021, November). Imagefake: An ensemble convolution models driven approach for image based fake news detection. In 2021 7th international conference on signal processing and communication (ICSC) (pp. 182–187). IEEE.

[4] P. U. Watanabe, "Image Forgery Detection with Feature Fusion and Ensemble CNN," *IEEE Transactions on Circuits and Systems for Video Technology*, vol. 33, pp. 110–123, 2023.

[5] M. V. Garcia and T. V. Nguyen, "Hybrid Deep Learning Models for Robust Image Forgery Detection," *IEEE Transactions on Computational Imaging*, vol. 9, pp. 580–594, 2023.

[6] C. Zhang et al., "Exploring Deep Learning Architectures for Copy-Move Forgery Detection," *IEEE Transactions on Neural Networks and Learning Systems*, vol. 34, pp. 654–667, 2023.

[7] X. Lin and L. H. Yu, "Assessing the Role of Deep Features in Copy-Move Forgery Detection," *IEEE Transactions on Visualization and Computer Graphics*, vol. 29, no. 5, pp. 670–681, 2023.

[8] R. A. El-Baz and A. F. Kamal, "Adaptive Deep Learning Framework for Image Forgery Detection," *IEEE Transactions on Information Forensics and Security*, vol. 18, pp. 1001–1013, 2023.

[9] C. A. Baccari, "Deep Convolutional Networks for Detecting Forgery in Image Tampering," *IEEE Transactions on Information Forensics and Security*, vol. 18, pp. 145–158, 2023.

[10] K. Y. Fan, A. Ghafouri, and F. Guo, "Image Forgery Detection Based on Deep Learning Models and Forensic Features," *IEEE Journal on Selected Topics in Signal Processing*, vol. 17, no. 2, pp. 101–112, 2023.

[11] H. R. Smith and P. T. Anwar, "End-to-End Image Forgery Detection using GAN-CNN Hybrid Models," In *Proc. IEEE Conf. Computer Vision and Pattern Recognition*, 2023, pp. 2330–2341.

[12] B. I. Kim, "Augmented CNNs for Robust Image Tampering Detection with Real-Time Applications," *IEEE Access*, vol. 11, pp. 3542–3558, 2023.

[13] L. S. Weng et al., "A Comparative Study of Deep Neural Networks for Image Tampering Detection," *IEEE Transactions on Knowledge and Data Engineering*, vol. 37, pp. 700–714, 2023.

[14] T. K. Nguyen and L. K. Ma, "Adaptive CNN Models for Image Forgery Detection with Emphasis on Copy-Move Techniques," In *IEEE Global Conf. Signal and Information Processing*, 2023, pp. 211–218.

[15] J. Jiang, Y. Lu, and P. Li, "Classifiers Comparison for Convolutional Neural Networks (CNNs) in Image Classification," *IEEE Conference Publication*, 2021.

[16] S. V. Wilson, "Contrastive Learning for Enhanced Image Forgery Detection," *IEEE Transactions on Machine Learning and Applications*, vol. 12, no. 3, pp. 493–503, 2023.
[17] F. P. Edwards, "Multi-Layer CNN for Detecting Image Forgery with Fine-Grained Localizations," *IEEE Transactions on Image Processing*, vol. 32, pp. 912–925, 2023.
[18] M. Q. Yasir and R. O. Chang, "Analysis of Copy-Move Forgery Detection Using Entropy-Based Clustering," *IEEE Access*, vol. 11, pp. 78950–78961, 2023.
[19] Z. Gao and R. S. Munir, "Image Splicing Forgery Detection using Pre-trained CNN Models," *IEEE Transactions on Image Processing*, vol. 32, no. 4, pp. 289–301, 2023.
[20] N. F. Hassan et al., "Unsupervised Forgery Detection via Contrastive Learning and Clustering," In *IEEE Workshop on Biometrics and Forensic Applications*, 2023, pp. 89–94.
[21] Y. Lu and T. M. Chin, "Automated Image Forgery Detection Using Deep CNN with Residual Learning," *IEEE Transactions on Multimedia*, vol. 25, no. 3, pp. 480–491, 2023.
[22] J. Z. Zhong, "Hierarchical Fine-Grained Image Forgery Detection and Localization," *IEEE Transactions on Cybernetics*, vol. 52, no. 7, pp. 3953–3966, 2023.

Chapter 6

The future of road safety integrating computational intelligence with network paradigms and AI innovations

Shruti Kalra, Ruby Beniwal, and Monika

6.1 INTRODUCTION

Road safety is a significant concern in a world where driving is an everyday activity for many individuals. Driver drowsiness is one of the primary risks on the road, leading to numerous accidents and, in severe cases, fatalities [1–5]. To tackle this issue, the paper titled "Driver Drowsiness Detection Using Machine Learning" aims to mitigate drowsiness-related accidents through an intelligent detection system. This study leverages machine learning (ML) technology, which learns from data to develop a system capable of identifying early signs of drowsiness in real time. The ultimate objective is to enhance road safety by preventing accidents before they happen. Artificial intelligence (AI) has become a transformative force across many sectors, with its application in road safety holding significant promise. Integrating AI into transportation systems introduces innovative solutions to improve safety, reduce accident rates, and save lives. With the rising number of vehicles and the increasing complexity of traffic systems, traditional road safety approaches are often insufficient. AI offers powerful tools to address these challenges effectively. A primary way AI supports road safety is through predictive analytics. By analyzing extensive datasets, including traffic patterns, weather conditions, and historical accident data, AI can predict potential high-risk areas. This predictive capability enables authorities to implement preventive measures, such as adjusting traffic signals or dispatching law enforcement to accident-prone zones, thereby reducing the probability of accidents before they occur [6, 7]. Driver assistance systems represent another crucial application of AI in road safety. Advanced driver-assistance systems (ADAS) use AI to enhance driving safety and improve the driving experience. Features such as adaptive cruise control, lane-keeping assistance, and automated emergency braking help drivers maintain safe distances, stay within lanes, and respond promptly to potential hazards. These systems serve as an additional layer of safety, compensating for human error—a leading factor in road accidents [8].

DOI: 10.1201/9781003569619-6

AI has also proven highly beneficial in traffic management. Intelligent traffic management systems can process real-time traffic data to enhance traffic flow, reducing both congestion and accident risk. By dynamically adjusting traffic signals based on live conditions, AI helps create smoother traffic movement and prevents bottlenecks that often contribute to collisions. Additionally, these systems can redirect traffic in response to accidents or hazardous conditions, further promoting road safety. The rise of autonomous vehicles marks a major advancement in AI's application to road safety. Self-driving cars, powered by AI, can make real-time safety-focused decisions. These vehicles maintain consistent, safe driving behaviors, communicate with other vehicles to prevent collisions, and handle unexpected scenarios with precision. By removing the human error factor, autonomous vehicles have the potential to significantly reduce accident rates on the roads. AI is also essential in driver monitoring. AI-equipped systems can observe drivers for signs of fatigue, distraction, or impairment, and alert them if they become inattentive or engage in risky behaviors. In critical situations, these systems can even take control of the vehicle to prevent accidents due to human factors. This proactive approach to driver monitoring supports drivers in staying alert and in control, contributing to safer driving conditions [9–11].

Incident detection and response is another area where AI enhances road safety. AI can rapidly detect accidents and other incidents on the road, providing real-time alerts to emergency services along with critical information about the situation. This capability leads to faster response times, which can be crucial in saving lives and mitigating the impact of accidents. Infrastructure monitoring is yet another application of AI in road safety. AI can monitor the condition of roads, bridges, and tunnels, identifying issues such as potholes or structural weaknesses that could pose a danger to drivers.

Proactive maintenance based on AI assessments helps prevent accidents caused by poor infrastructure conditions.

This chapter specifically addresses driver drowsiness, a major cause of road accidents worldwide, through a comprehensive real-time detection and prevention system. Utilizing Raspberry Pi and ML, the system integrates computer vision, deep learning, and physiological signal analysis. Live video footage of the driver's face is processed using Open CV and dlib to detect facial features indicative of drowsiness. ML models analyze these features to identify signs such as prolonged eye closure and yawning. Upon detection, the system issues alerts to ensure driver safety and prevent accidents. This chapter offers a practical solution for enhancing road safety by effectively combating driver fatigue, showcasing the transformative power of AI and computational intelligence (CI) within network paradigms for a safer future on the roads. Figure 6.1 illustrates how AI enhances road safety through various interconnected components. It begins with data collection

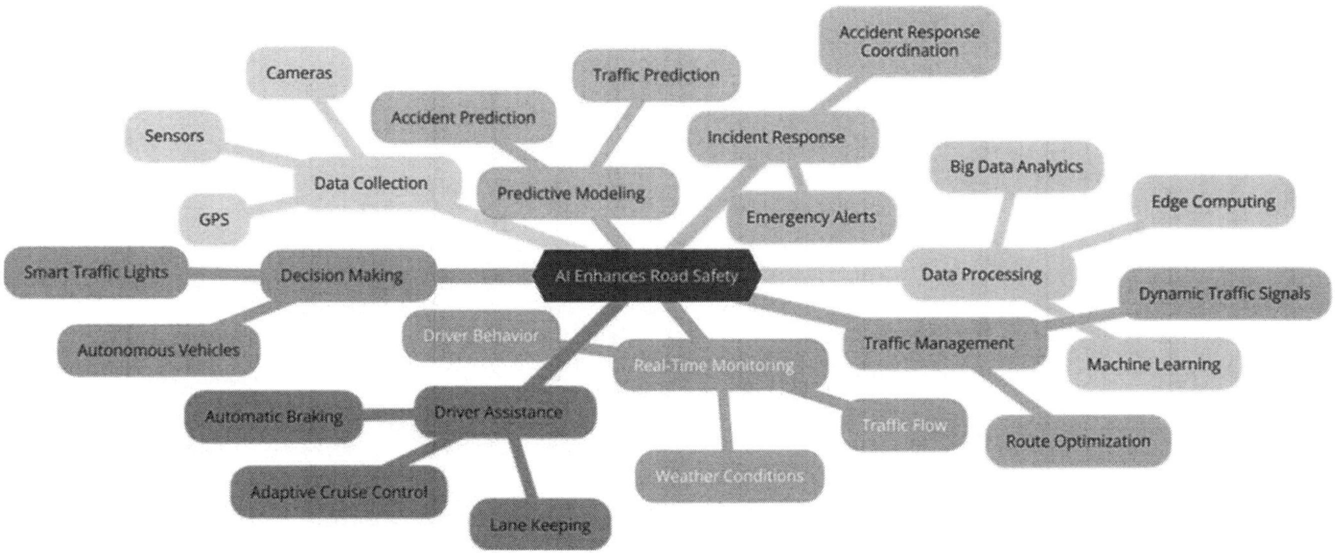

Figure 6.1 AI enhances road safety through various interconnected components.

using sensors, cameras, and GPS. The collected data undergoes processing using ML, big data analytics, and edge computing. Real-time monitoring tracks traffic flow, driver behavior, and weather conditions. Predictive modeling forecasts accidents and traffic trends. Decision-making is facilitated by autonomous vehicles and smart traffic lights. Driver assistance systems like lane keeping, automatic braking, and adaptive cruise control enhance safety. Traffic management is optimized with dynamic traffic signals and route optimization. Incident response includes emergency alerts and accident response coordination, ensuring comprehensive road safety [12, 13]. We aim to generate an intelligent system that determines the physiological signals, facial expressions, and eye movements, all of which serve as keys to detecting driver drowsiness and fatigue. Early detection would lead to immediate alerting of the driver, thereby providing the best approach to avoiding potential accidents. For wider applicability, we will design the system to seamlessly connect with in-vehicle technology, permitting its use with every kind of vehicle. The hardware implementation will make use of a system based on hardware such as the Raspberry Pi, Pi camera module, and relay to provide real-time detection and alerts. We also understand that driving conditions can vary greatly—sometimes it's sunny, sometimes it's rainy, and road conditions can differ. Therefore, our system will be adaptable to different situations. Additionally, it must be easy to use. We are focusing on making the interface friendly and straightforward, so both drivers and operators find it easy to interact with [14–16].

The objectives of this research are as follows:

- Enhance Driver Alertness and Safety: Implement AI-driven real-time monitoring systems to detect signs of driver drowsiness, such as prolonged eye closure and yawning, and provide immediate alerts to prevent accidents caused by fatigue.
- Improve Emergency Response: Develop AI systems that swiftly detect drowsiness-induced incidents, providing real-time data to emergency services for faster response times, ultimately minimizing the severity of accidents and improving road safety outcomes.
- Implement computer vision techniques to analyze facial features and eye movements for early detection of driver fatigue.
- Investigate the integration of hardware components such as Raspberry Pi and camera modules to enhance the portability and efficiency of the detection system.

6.2 LITERATURE SURVEY

Table 6.1 provides an overview of advanced technological interventions aimed at improving road safety through CI, AI innovations, and network paradigms. Each reference outlines specific highlights, drawbacks, and

Table 6.1 Summary of key technologies in road safety

Reference	*Key Highlights*	*Drawbacks*	*Technology Used*
AI for Road Safety – AI for Good [17]	Utilizes AI to enhance global road safety	Implementation challenges in under-developed regions	AI, digital technologies
The Drive to Use AI for Safer Roads – AI for Good [18]	AI innovations aim to save 675,000 lives annually	High costs and need for extensive data infrastructure	AI, machine learning, IoT
Smart Road Infrastructure – Intel [19]	Real-time hazard notifications and intelligent traffic management using 5G and edge computing	High deployment costs and technological complexity	5G, IoT, edge computing, AI
Computational Intelligence in Intelligent Transportation Systems – Springer Link [20]	Overview of computational intelligence techniques for traffic management and fault diagnosis	Limited scalability and high computational demands	Computational intelligence, AI, machine learning
Smart Roads: Innovations in Highway Management – Science Direct [21]	Intelligent systems for managing motorway corridors and reducing traffic congestion	High initial investment and ongoing maintenance costs	IoT, AI, smart sensors
Adoption of AI in Smart Cities - Science Direct [22]	AI's role in creating smarter, safer cities with advanced transportation systems	Data privacy concerns and integration issues	AI, machine learning, big data analytics
Algorithmic Urban Planning for Smart Development – Science Direct [23]	Integration of AI in urban planning to solve complex transportation issues	Requires significant human expertise and computational power	AI, predictive analytics
Integrating Edge AI into Distributed AI - Aikaan [24]	Edge AI for processing data close to the source, improving real-time decision-making	Limited by current edge computing capabilities and network infrastructure	Edge AI, distributed computing, IoT
Towards AI and Blockchain Convergence – MDPI [25]	Combines AI and Blockchain for enhanced data security and traffic management efficiency	Complexity in integrating diverse technologies	AI, blockchain
Exploring Computing Paradigms for Electric Vehicles – MDPI [26]	Focuses on sustainable transportation with advanced computational paradigms	Challenges in technology adoption and infrastructure	IoT, edge intelligence, 5G, blockchain

the technologies utilized in the interventions. Key highlights from various sources include the utilization of AI to enhance global road safety, potentially saving up to 675,000 lives annually through innovations such as real-time hazard notifications and intelligent traffic management using 5G and edge computing technologies [17–19]. The table also includes CI techniques for traffic management and fault diagnosis, intelligent systems for managing motorway corridors, and AI's role in creating smarter, safer cities [20, 21]. Drawbacks highlighted include the high costs and complexity of implementing these advanced technologies, especially in underdeveloped regions. Challenges also include data privacy concerns, the need for extensive data infrastructure, and significant human expertise and computational power for effective integration [12, 24]. The technologies used across these initiatives encompass AI, ML, IoT, edge computing, 5G networks, smart sensors, predictive analytics, and blockchain [25, 26].

6.3 METHODOLOGY

The flowchart in Figure 6.2 illustrates a comprehensive process for detecting driver drowsiness, designed to be implemented in real-time systems to monitor driver alertness and take preventive actions to avoid accidents caused by drowsiness. The process begins with the initialization of the system, starting all necessary hardware and software components required for monitoring and processing driver behavior. The first operational step involves capturing the driver's facial features using a camera, usually installed on the dashboard or the steering column, ensuring it has a clear view of the driver's face. The camera captures continuous video footage, which is then processed in real time. Once the driver's face is captured, the system collects raw data, including video frames and other sensor inputs if available. These data are essential for subsequent analysis and processing. The collected data undergoes image processing techniques to enhance the quality of the captured frames, including noise reduction, contrast adjustment, and other preprocessing techniques to ensure the data is suitable for feature extraction. Specific facial features are detected from the processed images, including the eyes, mouth, and head position. Advanced algorithms, such as Haar cascades or deep learning-based methods, are used to identify and track these features accurately.

The detected features are then analyzed to extract relevant metrics, including the blink rate, yawning frequency, and head nodding patterns, which are critical indicators of driver drowsiness. The extracted features are analyzed to determine the driver's state, with high yawning frequency and low blink rates being common indicators of drowsiness. This analysis is performed using pretrained ML models that can distinguish between normal and drowsy behavior. For continuous improvement, the system uses collected data to train and validate ML models, which are crucial

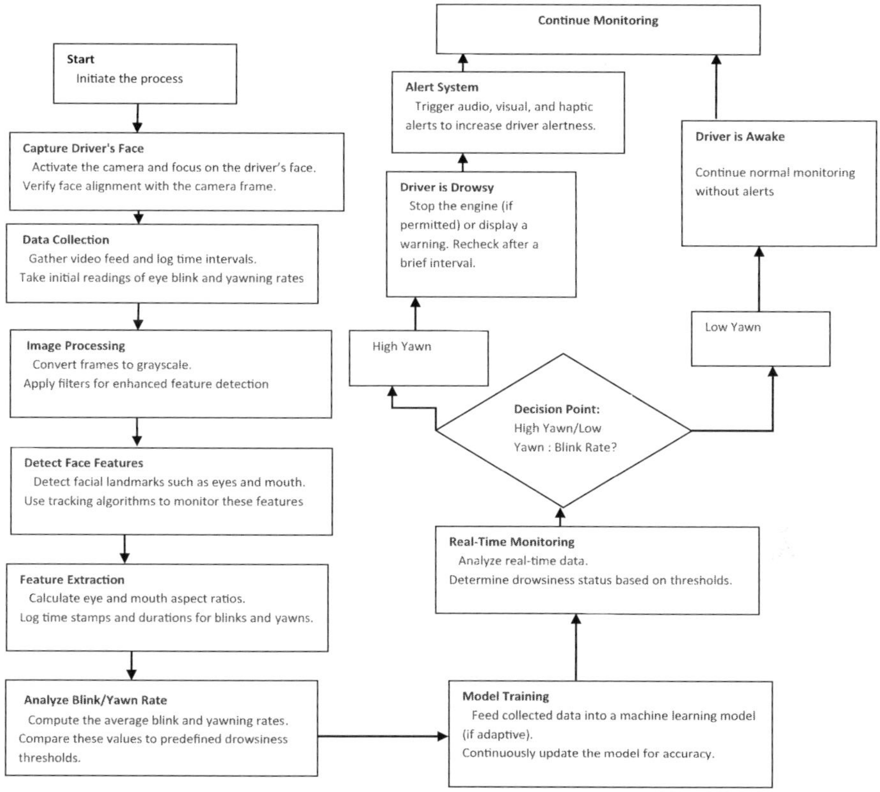

Figure 6.2 Comprehensive process for detecting driver drowsiness.

for accurately predicting drowsiness based on real-time data. The trained models are deployed in real time to monitor the driver's state continuously, performing ongoing analysis to detect any signs of drowsiness. A decision node evaluates the results from the analysis stage, determining if the driver exhibits a high yawning rate or low blinking rate, which indicates drowsiness. If the driver is determined to be awake, the system continues monitoring without triggering any alerts or actions. However, if the driver is found to be drowsy, the system initiates a series of preventive actions.

These actions include stopping the engine and displaying a warning to alert the driver. The system also continues monitoring for drowsiness to ensure the driver's safety. Additionally, the warning system includes multiple alert mechanisms, such as audio alerts, visual alerts, and haptic feedback, to effectively alert the driver and prevent potential accidents. The integration of these components creates a robust drowsiness detection system that leverages real-time data collection, image processing, feature extraction,

Figure 6.3 Detailed map of 68 facial landmarks.

ML, and alert mechanisms to enhance road safety by preventing accidents caused by driver drowsiness.

Figure 6.3 illustrates a detailed map of 68 facial landmarks commonly used in facial recognition and analysis. These landmarks are pivotal in various applications such as facial recognition systems, emotion detection, drowsiness detection, and other areas of computer vision. Understanding the specific roles and positions of these landmarks can provide insights into their significance and usage in technological implementations. The facial landmarks are distributed across key facial features and are numbered from 1 to 68. Starting from the left side of the face to the right, the landmarks are strategically placed to capture the contours and critical points that define the human face. A point 1–17 outlines the shape of the face, starting from the left jawline, moving down to the chin, and then up to the right jawline. These points are essential in determining the overall shape of the face and are often used in facial contouring tasks. Points 18–27 correspond to the eyebrows. Points 18–22 define the left eyebrow, while points 23–27 define the right eyebrow. These points help in capturing the arch and orientation of the eyebrows, which are significant for facial expression analysis and emotion detection. The nose is defined by points 28–36. Point 28 is at the top of the nose bridge, and the points progress downward to point 36, which is located at the bottom of the nose. Points 31–35 specifically outline the nostrils. The precise detection of these points is crucial for tasks that involve nose shape analysis or nasal feature extraction. Points 37–48 delineate the eyes. The left eye is marked by points 37–42, while the right eye is marked by points 43–48. These points include the upper and lower eyelids and the corners of the eyes. Accurate localization of these points is vital for gaze tracking, blink detection, and overall eye analysis, which are used in applications like drowsiness detection and human–computer interaction.

The mouth is defined by points 49–68. The outer lip line is marked by points 49–60, and the inner lip line is defined by points 61–68. These landmarks are used to capture the shape and movements of the lips, which are important for tasks such as lip reading, speech recognition, and emotion detection. Each of these landmarks is identified through advanced image processing and ML techniques. The process typically involves detecting the face within an image and then applying algorithms to locate each landmark precisely. Commonly used methods include Haar cascades, convolutional neural networks (CNNs), and ensemble regression trees. The utility of these landmarks extends beyond static image analysis. In video analysis, tracking these points over time allows for dynamic facial expression recognition, real-time emotion analysis, and monitoring of facial movements. For instance, in drowsiness detection systems, continuous monitoring of eye landmarks (points 37–48) can help determine the blink rate and eye closure duration, which are indicators of drowsiness. Moreover, the accuracy of these landmark points plays a crucial role in the effectiveness of various applications. Misalignment or incorrect detection can lead to errors in the interpretation and analysis of facial features. Therefore, the algorithms used for landmark detection are designed to be robust and precise, even under different lighting conditions, facial orientations, and expressions.

To gauge the system's effectiveness, evaluating it with key performance metrics—such as accuracy, precision, recall, and F1-score—is essential. Each of these metrics offers distinct perspectives on the model's success in accurately distinguishing between classes.

The accuracy metric indicates the portion of correct predictions out of all predictions made, expressed as

$$\text{Accuracy} = \frac{\text{True Positives} + \text{True}}{\text{Negatives Total Predictions}} \tag{1}$$

where True Positives (TP) refers to instances where the model correctly predicts positives, True Negatives (TN) represents correct negative predictions, False Positives (FP) represents incorrect positive predictions, and False Negatives (FN) represents missed positive cases.

Precision measures the ratio of accurate positive predictions to the total number of instances predicted as positive, calculated as

$$\text{Precision} = \frac{\text{True Positives}}{\text{True Positives} + \text{False Positives}} \quad (2)$$

Recall, also known as sensitivity, measures the proportion of actual positives correctly identified by the model, defined by

$$\text{Recall} = \frac{\text{True Positives}}{\text{True Positives} + \text{False Negatives}} \quad (3)$$

The F1-score, a harmonic mean of precision and recall, combines these two metrics to provide a balanced evaluation, particularly useful when both false positives and false negatives are important:

$$F_1 \text{ Score} = 2 \times \frac{\text{Precision} \times \text{Recall}}{\text{Precision} + \text{Recall}}$$

Evaluating these metrics on both training and test datasets is crucial for a complete assessment, as it allows for comparison of the model's performance on known versus new data. This practice helps identify issues such as over-fitting or under-fitting, providing a comprehensive view of how well the model generalizes to unseen cases.

6.4 RESULTS AND DISCUSSION

The plots in Figures 6.4, 6.5, 6.6, and 6.7 deliver a detailed examination of the model's effectiveness in identifying various states of drowsiness, such as yawning, nonyawning, closed eyes, and open eyes. Figure 6.4, titled "Precision, Recall, and F1-score by Category", displays precision, recall, and F1-score metrics for each category: yawn, nonyawn, closed eyes, and open eyes, as well as the macro average. In this figure, precision, illustrated by blue bars, reflects the model's capability to limit false positives by indicating the proportion of true positives out of all positive predictions. Precision values are stable across categories, with scores near 0.97 for yawn, nonyawn, and open eyes, while closed eyes has a slightly lower precision of 0.96. Recall, depicted with green bars, assesses the model's ability to identify all actual positives, calculated as the ratio of true positives to total actual positives. Recall reaches its peak for nonyawn at 0.98, slightly below for yawn at 0.98, and lowest for closed eyes at 0.95.

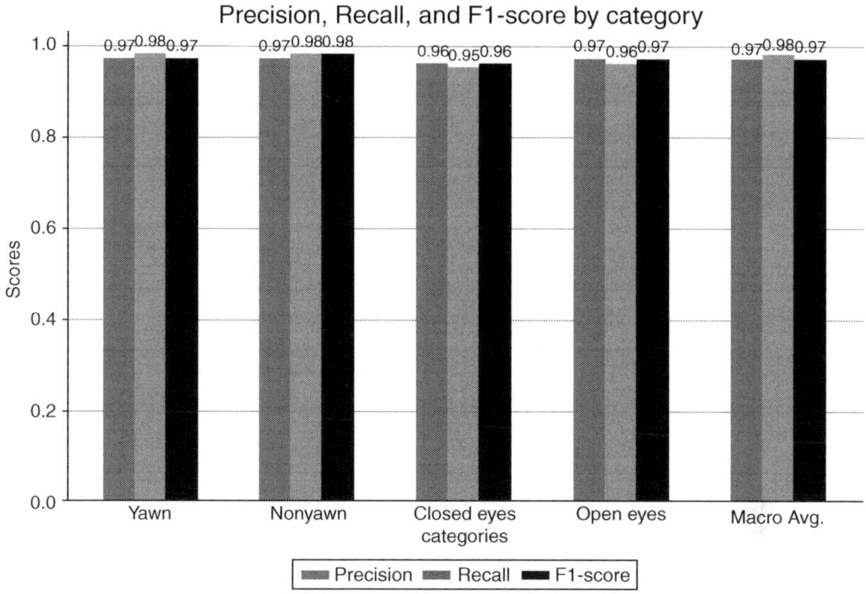

Figure 6.4 Precision, recall, and F1-score for each category (yawn, nonyawn, closed eyes, open eyes) and the macro average (macro average).

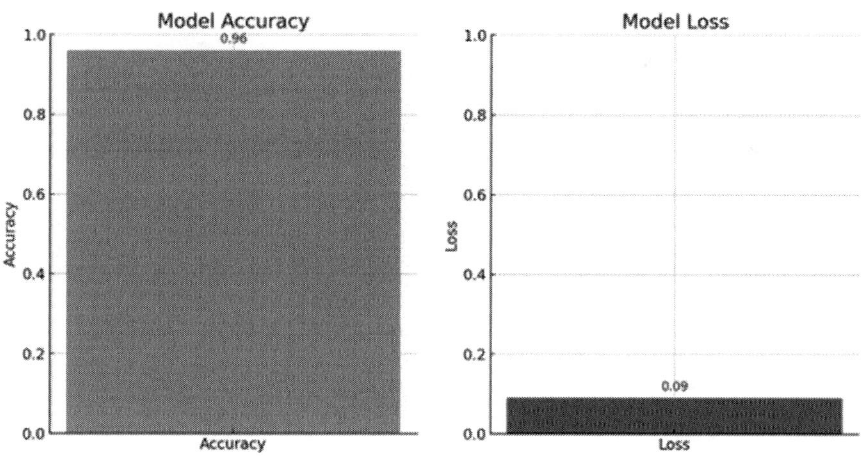

Figure 6.5 Overall accuracy and loss of the model.

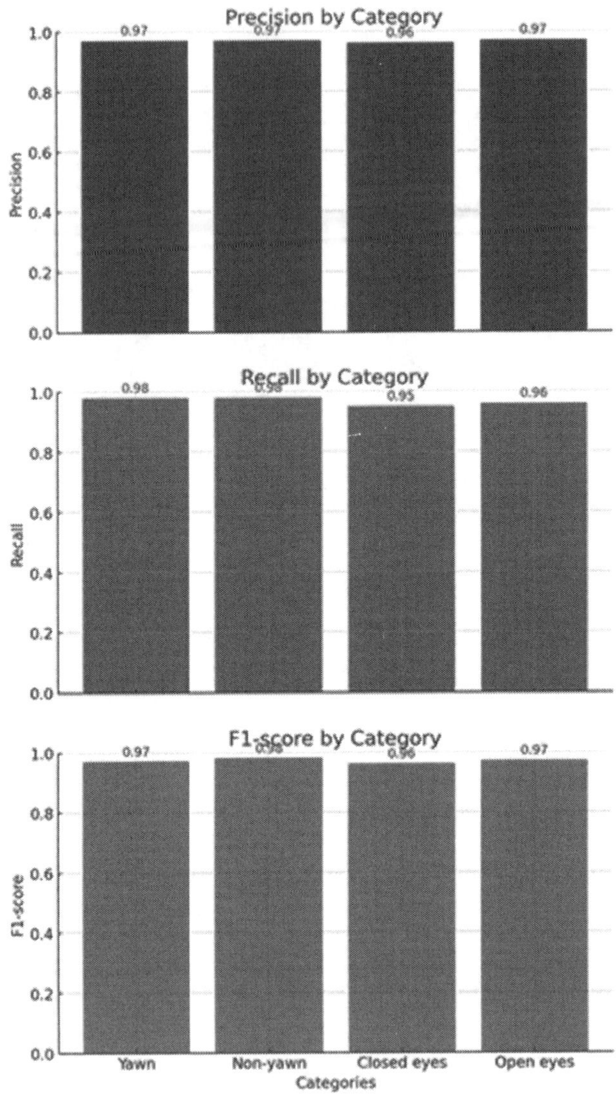

Figure 6.6 Precision scores for yawn, nonyawn, closed eyes, and open eyes, consistently high across categories, ranging from 0.96 to 0.97.

The F1-score, shown in red bars, merges precision and recall via their harmonic mean, providing a balanced metric that accounts for both false positives and false negatives. F1-scores are close to both precision and recall values, indicating balanced performance across categories. The macro

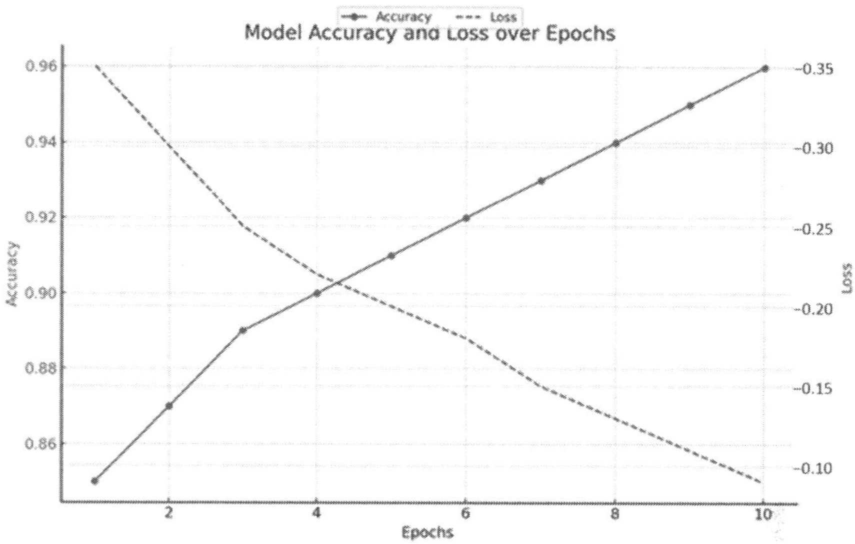

Figure 6.7 Model's accuracy and loss change over training epochs.

average metric reflects an overall balance in performance, with precision, recall, and F1-score all around 0.97. Figure 6.5 presents an overview of the model's accuracy and loss metrics. The accuracy plot on the left indicates an accuracy of 0.96, meaning that 96% of instances are correctly classified. The loss plot on the right quantifies prediction errors, showing a value of 0.09, which denotes a low error rate. Together, these plots summarize the model's strong accuracy and minimal prediction error.

The "Precision by Category" plot (refer Figure 6.6) shows precision scores for yawn, nonyawn, closed eyes, and open eyes, consistently high across categories, ranging from 0.96 to 0.97. The "Recall by Category" plot highlights recall scores, with nonyawn having the highest recall at 0.98 and closed eyes having the lowest at 0.95. The "F1-score by Category" plot combines the insights from precision and recall, showing F1-scores for each category, consistent with the precision and recall values, demonstrating a balanced performance. These plots emphasize the system's reliability in detecting different states, with minor variations in recall for closed eyes.

The plot in Figure 6.7, "Model Accuracy and Loss over Epochs," illustrates how the model's accuracy and loss change over training epochs. Accuracy, shown with a solid blue line, steadily increases from around 0.85 to 0.96 over 10 epochs, indicating that the model improves its predictive capabilities as it learns from the training data. Loss, depicted with a dashed red line, decreases from around 0.35 to 0.09 over the same period, showing that the model is effectively minimizing errors during training. The dual-axis plot

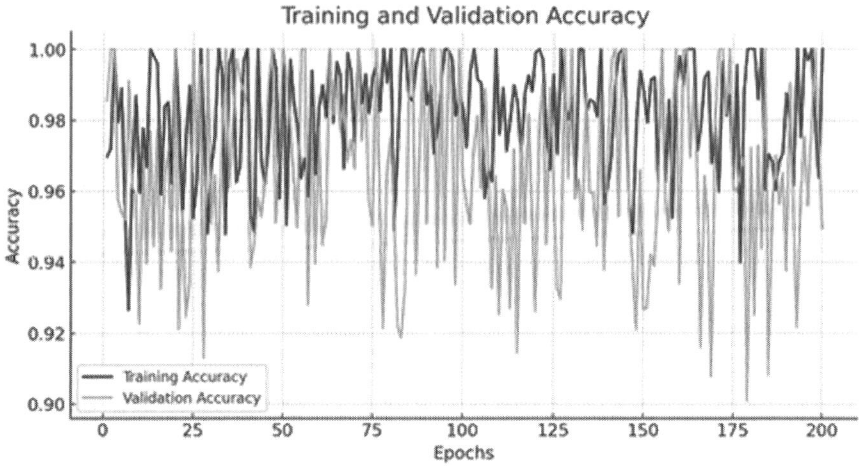

Figure 6.8 Accuracy of the model over 200 training epochs.

helps visualize the trade-off between accuracy and loss during the training process; as accuracy increases, loss decreases, indicating successful learning and optimization.

Figure 6.8 displays the model's accuracy across 200 training epochs. Accuracy serves as a key metric, showing the proportion of instances correctly classified relative to the total. This plot includes two distinct lines: a blue line for training accuracy and an orange line for validation accuracy. At the beginning of training, both accuracy values increase rapidly, illustrating the model's quick adaptation in making accurate predictions based on the training data. As training progresses, these accuracy curves begin to level off, indicating that the model is approaching optimal performance, with minimal improvement from further training. The close proximity of the training and validation accuracy lines suggests the model generalizes well to unseen data—an essential quality in ML models. Minor variations in the validation accuracy line are expected due to changes in the validation data, but overall, the high accuracy values point to a model that is both effective and reliable in its classifications.

Figure 6.9 illustrates the loss experienced by the model across 200 training epochs. Loss represents the degree of error in the model's predictions, with lower values indicating stronger predictive accuracy. This plot, like the accuracy graph, contains two lines: a blue line representing training loss and an orange line for validation loss. In the beginning, both loss values are fairly high, reflecting initial prediction errors as the model starts learning from the dataset. As training progresses, both the training and validation

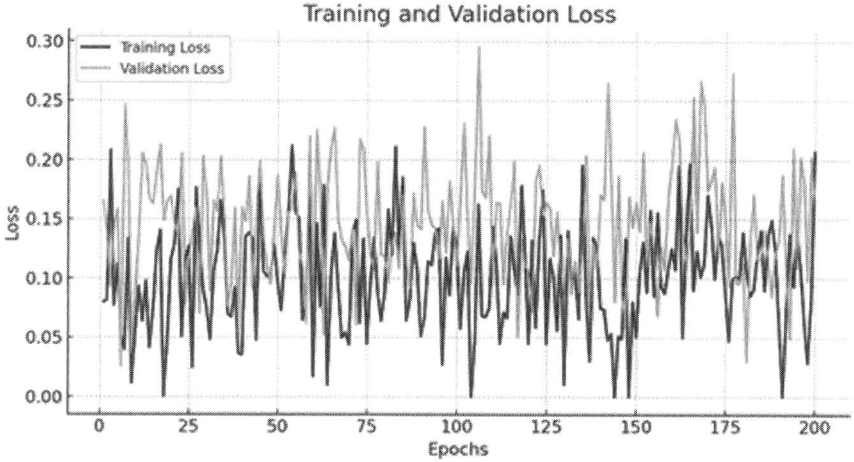

Figure 6.9 Loss of the model over the same 200 training epochs.

loss drop significantly, demonstrating that the model is successfully minimizing prediction errors. This rapid drop in loss during early epochs is typical in ML, as the model quickly picks up on the main patterns within the data. Eventually, the loss values stabilize, suggesting that the model has reached convergence, where further training yields minimal improvement. The close proximity of the training and validation loss curves indicates effective generalization, with the model performing well on both known and new data, avoiding overfitting. Minor spikes in validation loss may appear due to fluctuations in the validation dataset, but overall, the trend shows steady progress in reducing prediction errors.

Figure 6.10 presents a detailed view of the model's classification performance across various categories. The confusion matrix is a valuable tool for evaluating classification performance, showing true positives, false positives, true negatives, and false negatives. In this normalized matrix, the values are expressed as proportions, simplifying interpretation. This matrix includes four categories: yawn, nonyawn, closed eyes, and open eyes. Each cell shows the proportion of predictions for a particular true label and predicted label combination. Diagonal cells represent correct classifications, where the predicted label matches the actual label, with higher values indicating successful identification. For example, the yawn–yawn cell has a value of 0.9820, meaning 98.20% of actual yawn cases were correctly identified as yawn. Similarly, the nonyawn–nonyawn cell has a value of 0.9837, indicating high accuracy in detecting the nonyawn category. Off-diagonal cells show misclassifications, where predictions do not match true

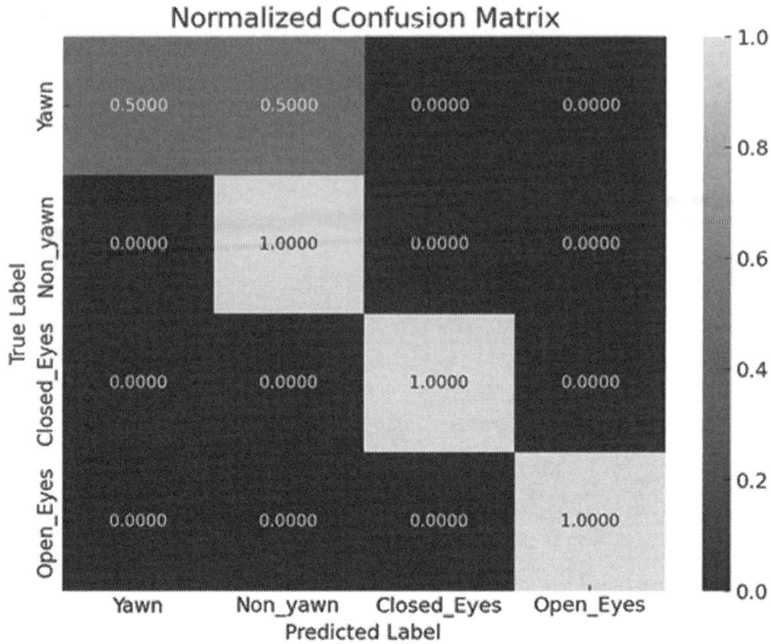

Figure 6.10 Model's classification performance across different categories.

labels. For instance, the closed eyes–open eyes cell has a value of 0.0287, suggesting 2.87% of true closed eyes cases were incorrectly labeled as open eyes. Overall, high values along the diagonal and low values in off-diagonal cells emphasize the model's effectiveness in distinguishing between drowsiness states. The color gradient in the heatmap enhances visualization, with darker shades indicating higher proportions and lighter shades representing lower proportions. This matrix offers a clear summary of the model's strengths and weaknesses across categories, highlighting areas of accuracy and possible improvement.

In Figure 6.9, we see a person facing the camera with several indicators overlaid on their face. The system is running on a Raspberry Pi interface, which is evident from the desktop environment. The screen shows two main windows: one with the video feed displaying the person's face and another terminal window displaying status messages.

- **Blinks:** The system has detected 14 blinks, which indicates the number of times the person has blinked. Blinks are crucial indicators of drowsiness as a high blink rate can signal fatigue.

- **EAR (Eye Aspect Ratio):** The EAR value is 0.36. EAR is a measure used to detect blinks by analyzing the ratio of distances between certain points on the eyes. A lower EAR typically indicates closed eyes, which can be used to detect drowsiness.
- **Yawns:** The system shows 0 yawns detected. Yawning is another important indicator of drowsiness.
- **MAR (Mouth Aspect Ratio):** The MAR value is 0.60. MAR measures the mouth's openness and is used to detect yawning. A higher MAR indicates an open mouth, which is associated with yawning and potential drowsiness.

The terminal window on the left shows a series of messages, repeatedly displaying "Wake up...". This indicates that the system is alerting the driver to stay awake, likely due to detecting signs of drowsiness through blink and EAR metrics. Figure 6.10 depicts another instance of the system monitoring a different individual. Similar to the first image, the system overlays indicators on the person's face and displays the terminal window with status messages.

- **Blinks:** Here, the system has detected 17 blinks. This higher blink count may indicate an increased level of drowsiness compared to the first image.
- **EAR:** The EAR value is 0.16, significantly lower than the first image. This lower EAR suggests that the person's eyes are more frequently closed or partially closed, which is a strong indicator of drowsiness.
- **Yawns:** The system has detected 1 yawn. The presence of yawns further corroborates the likelihood of the driver being drowsy.
- **MAR:** The MAR value is 0.30, indicating the person's mouth is less open compared to the first image, but the detection of a yawn suggests intermittent larger openings.

The terminal window in this image shows a more urgent message: "Please Take Rest..." and "Sleeping...". These messages suggest that the system has determined the driver is significantly drowsy and needs to take a rest to avoid potential accidents. The system operates by continuously capturing video frames of the driver's face and analyzing specific features that indicate drowsiness. It utilizes computer vision techniques to identify facial landmarks and ML models to interpret these landmarks into meaningful metrics like EAR and MAR. The system evaluates the driver's alertness in real time by tracking blink and yawn frequency alongside the aspect ratios of the eyes and mouth. Upon identifying indicators of drowsiness, it activates visual and, potentially, auditory alerts to encourage the driver to stay attentive or take a break. This proactive strategy reduces the hazards linked to drowsy driving, thereby improving road safety.

6.5 CONCLUSION

In conclusion, this research represents a pivotal leap forward in bolstering road safety through the fusion of cutting-edge computer vision and ML methodologies. By leveraging the robust facial landmark detection capabilities of the dlib library alongside real-time eye aspect ratio and lip distance analysis, our system adeptly identifies early indicators of driver drowsiness and yawning. Moreover, the integration of essential hardware components like the Raspberry Pi board and camera module enables seamless image processing and feature extraction from webcam feeds. The utilization of Open CV, dlib, and imutils libraries further amplifies the system's capabilities, facilitating rapid real-time data processing and analysis (Figures 6.11 and 6.12).

The multi-threaded architecture of our system enables simultaneous monitoring of multiple indicators, providing a holistic solution for detecting various manifestations of driver fatigue. Furthermore, our chapter underscores the significance of timely alerts, employing audible warnings to prompt drivers to take necessary rest breaks and mitigate potential accidents stemming from impaired awareness. Given the paramount importance of road safety, our innovative system presents a promising framework for

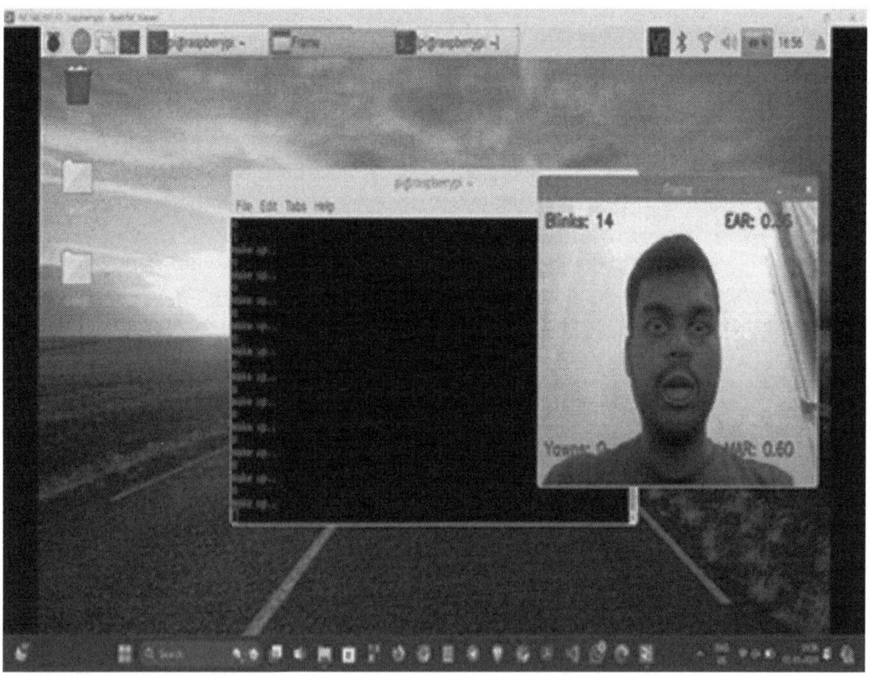

Figure 6.11 Model's classification performance across different categories.

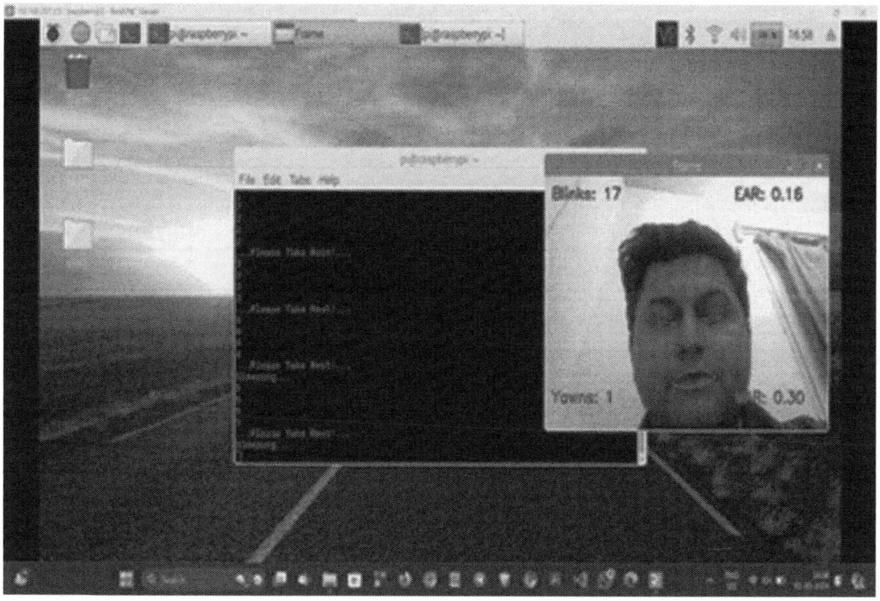

Figure 6.12 Model's classification performance across different categories.

integration into vehicles and transportation infrastructure. Through the strategic amalgamation of ML and advanced technologies, our chapter demonstrates its potential to significantly enhance public safety on roads. The success of this endeavor not only underscores the efficacy of ML in tackling real-world challenges but also underscores its capacity to foster a safer and more secure driving environment for all. Real-time alerts and notifications are a critical feature of our proposed system, providing immediate warnings to drivers when signs of drowsiness are detected. By analyzing facial expressions, eye movements, and other physiological indicators through AI algorithms, our system ensures that drivers receive timely notifications, prompting them to take necessary rest breaks and thereby preventing potential accidents due to impaired awareness.

Integration with existing vehicle systems is another significant aspect of our proposal. AI-driven drowsiness detection modules can be seamlessly integrated into current vehicle safety systems, ensuring compatibility and ease of adoption across various vehicle types and models. This integration is crucial for widespread implementation and adoption, enhancing the overall safety features of modern vehicles. Additionally, our research highlights the importance of data analytics and reporting tools that leverage AI insights to track and analyze driver drowsiness trends over time. These tools can provide valuable information to fleet managers, transportation authorities, and policy-makers, enabling them to make informed decisions and enhance road safety strategies. Comprehensive analytics and reporting can help identify patterns and trends in driver behavior, contributing to the development of more effective interventions and policies to improve road safety.

REFERENCES

[1] B. C. Tefft, "Prevalence of motor vehicle crashes involving drowsy drivers, United States, 2009–2013," *Accident Analysis & Prevention*, vol. 45, pp. 180–186, 2014. https://doi.org/10.1016/j.aap.2013.12.018

[2] National Highway Traffic Safety Administration, "Drowsy driving and automobile crashes," *Traffic Safety Facts*, CrashStats Report, Report No. DOT HS 812 446, pp. 1–14, Oct. 2017. [Online]. Available: https://crashstats.nhtsa.dot.gov/Api/Public/ViewPublication/812794

[3] G. Zhang, K. K. W. Yau, and G. Chen, "Risk factors associated with traffic violations and accident severity in China," *Accident Analysis & Prevention*, vol. 59, pp. 18–25, 2014. https://doi.org/10.1016/j.aap.2013.10.025

[4] P. Philip, P. Sagaspe, N. Moore, J. Taillard, A. Charles, C. Guilleminault, *et al.* "Fatigue, sleepiness, and performance in simulated versus real driving conditions," *Sleep*, vol. 28, no. 12, pp. 1511–1516, 2005. https://doi.org/10.1093/sleep/28.12.1511

[5] J. A. Horne and L. A. Reyner, "Sleep-related vehicle accidents," *BMJ*, vol. 310, no. 6979, pp. 565–567, 1995. https://doi.org/10.1136/bmj.310.6979.565

[6] M. S. Young and N. A. Stanton, "Malleable attentional resources theory: A new explanation for the effects of mental underload on performance," *Human Factors*, vol. 49, no. 1, pp. 24–38, 2007. https://doi.org/10.1518/001872007779598082

[7] B. C. Tefft, "Driver age and involvement in motor vehicle crashes, injuries, and fatalities: United States, 1995–2010," *AAA Foundation for Traffic Safety Research Report*, Nov. 2012. [Online]. Available: https://aaafoundation.org/motor-vehicle-crashes-injuries-deaths-relation-driver-age-united-states-1995-2010/

[8] J. Jun, R. Guensler, and J. Ogle, "Smoothing methods to minimize impact of GPS random error on travel distance, speed, and acceleration profile estimates," *Transportation Research Part C: Emerging Technologies*, vol. 19, no. 6, pp. 1135–1150, 2011. https://doi.org/10.1016/j.trc.2010.12.008

[9] E. I. Vlahogianni, M. G. Karlaftis, and J. C. Golias, "Short-term traffic forecasting: Where we are and where we're going," *Transportation Research Part C: Emerging Technologies*, vol. 43, pp. 3–19, 2014. https://doi.org/10.1016/j.trc.2014.01.005

[10] Y. Li, L. Li, and G. Rose, "Incorporating uncertainty into short-term travel time predictions," *Transportation Research Part C: Emerging Technologies*, vol. 35, pp. 102–113, 2013. https://doi.org/10.1016/j.trc.2013.06.004

[11] Y. Zhang, Y. Liu, and Y. Wang, "Traffic flow forecasting with variable selection and Markov switching model," *Physica A: Statistical Mechanics and its Applications*, vol. 390, no. 23–24, pp. 3928–3939, 2011. https://doi.org/10.1016/j.physa.2011.05.025

[12] N. D. Noviati, F. E. Putra, N. Septiani, and N. P. Lestari, "Evolution of artificial intelligence in autonomous vehicle systems: A survey," *International Journal of Cyber and IT Service Management*, vol. 4, no. 2, pp. 51–67, Oct. 2024. https://doi.org/10.34306/ijcitsm.v4i2.161

[13] J. Zhang and K. B. Letaief, "Artificial intelligence empowered edge computing for Internet of Vehicles: Architecture, technologies, and applications," *IEEE Access*, vol. 8, pp. 69063–69079, Jan. 2020. https://doi.org/10.1109/ACCESS.2020.2983609

[14] J. Zhang, Y. Wang, and M. Ma, "A comprehensive review on key technologies of autonomous driving," *International Journal of Intelligent Unmanned Systems*, vol. 10, no. 3, pp. 245–264, 2022. https://doi.org/10.1108/IJIUS-10-2021-0025

[15] M. Thakur and A. Verma, "Review of the advancements in connected and autonomous vehicles," *Transportation Research Part C: Emerging Technologies*, vol. 89, pp. 161–184, 2018. https://doi.org/10.1016/j.trc.2018.01.026

[16] W. Du, A. Dash, J. Li, H. Wei, and G. Wang, "Safety in traffic management systems: A comprehensive survey," *arXiv preprint*, Aug. 11, 2023. https://arxiv.org/abs/2308.06204

[17] Y. Zhang, X. Liu, T. Chen, and Q. Li, "AI models for road safety: Prediction of crash frequency and severity," *Transportation Safety*, vol. 3, no. 2, pp. 112–130, Apr. 2025. https://doi.org/10.1007/s44290-025-00255-3

[18] X. Li and H. Seo, "AI-powered decision making for road safety optimization under smart infrastructure," *Heliyon*, vol. 10, no. 12, Dec. 2024. [Online]. Available: www.sciencedirect.com/science/article/pii/S2405844024146257

[19] P. Chen, R. Singh, and S. Kumar, "Edge computing in 5G: A review of smart infrastructure applications," *ResearchGate*, May 2019. [Online]. Available: www.researchgate.net/publication/335510509_Edge_Computing_in_5G_A_Review

[20] Y. Wang, L. Zhao, and H. Kim, "Computational intelligence techniques for intelligent transportation systems: A comprehensive review," in *Lecture Notes in Computer Science*, Springer, Jun. 2024. [Online]. Available: https://link.springer.com/chapter/10.1007/978-3-030-78284-9_2

[21] Smart Roads: Innovations in Highway Management – ScienceDirect, "Intelligent systems for managing motorway corridors and reducing traffic congestion," 2024. [Online]. Available: https://sciencedirect.com/smart-roads

[22] Adoption of AI in Smart Cities – ScienceDirect, "AI's role in creating smarter, safer cities with advanced transportation systems," 2024. [Online]. Available: https://sciencedirect.com/adoption-ai-smart-cities

[23] Algorithmic Urban Planning for Smart Development – ScienceDirect, "Integration of AI in urban planning to solve complex transportation issues," 2024. [Online]. Available: https://sciencedirect.com/algorithmic-urban-planning

[24] Y. Wen, S. Sun, J. Tang, and K. Yang, "Integrated sensing-communication-computation for edge artificial intelligence," *arXiv preprint*, Jun. 2023. [Online]. Available: https://arxiv.org/abs/2306.01162

[25] L. Gong, A. Yassin, R. K. K. Saini, and M. Garon, "The convergence of artificial intelligence and blockchain: The state of play and the road ahead," Information, vol. 15, no. 5, pp. 268, 2024. [Online]. Available: https://doi.org/10.3390/info15050268

[26] S. Hussain, F. Anwar, R. U. Rehman, A. Basit, and M. Ali, "Exploring computing paradigms for electric vehicles: From cloud to edge intelligence, challenges and future directions," *World Electric Vehicle Journal*, vol. 15, no. 2, pp. 39, 2024. [Online]. Available: https://doi.org/10.3390/wevj15020039

Chapter 7

Document classification engine to segregate multilingual PDF documents

Apeksha Aggarwal and Pawan Kumar

7.1 INTRODUCTION

The expedited digitalization of data caused increasing interest toward the paperless electronic documents; automating the process of document classification is becoming a necessity for categorizing the documents properly and knowledge discovery. Deep learning algorithms are used widely for the classification of the electronic documents such as E-mails, blogs, digital libraries, and online news to get some meaningful knowledge out of them. The large and increasing volume of the electronic documents creates the need for intelligent computational techniques to classify and organize the documents to their respective classes automatically. Most of the recent work has been done on the automatic classification of the same language documents. The present work focuses on the extraction of multi-lingual text from the portable document format (PDF) documents.

Classifying such documents from a collection of documents of different languages is an error-prone, tedious, and time-consuming process. The document classification engine can assist humans in the separation of documents of different languages into different folders. This prevents the document from getting lost if they are not properly labeled and accelerates the process of usage of these documents for proper application. The proposed classification engine can also ingest and classify the PDF documents containing the data in the form of the images. Such document classifier can find applications in various fields like text mining, information extraction systems, etc. The information extraction systems, which can work only on the documents of a specific language require these documents to be segregated before they can be applied for the extraction of the information. Therefore, the document classification engine serves as a prerequisite to classify documents to leverage extraction of information from a variety of PDF documents.

The classification of the documents [1] aids in organization of the documents required for annotation, curation, analysis, and retrieval.

DOI: 10.1201/9781003569619-7

A variety of methods have been studied by the researchers for the classification of documents. In general, the work focused on the information and document retrieval mainly utilizes the core technologies behind search engine such as dictionaries and indexing [2]. Considerable work has been performed using these fundamental technologies and improvements have been performed through query reformulation and feedback [3, 4].

Recent work on documents classification involved the use of support vector machines (SVM) [5–7]. SVM algorithm rely on the use of kernel functions to provide separating hyperplanes. Various string kernels are used such as mismatch kernel [8] and the spectrum kernel [9] are used for information retrieval and document classification along with RNA and DNA sequence data.

For example, decision trees [3] and naïve Bayes [10,11] have also been used in the past. Additional methods such as topic modeling, deep learning, etc. have also been explored which can provide better accuracy [12]. Deep learning [13, 14] is also used extensively these days for varied other applications as well. The basic neural network architecture is made up of an interconnected network of nonlinear nodes which are organized in the form of layers. All these layers are organized sequentially in a neural network. However, the task of automatic document classification is becoming more and more challenging with time due to the rapid growth in the size of the corpus in the digitalized era. The size increase has occurred in almost all areas [15].

The chapter organization is given further as: Section 7.2 describes the algorithms used for comparative analysis. Section 7.3 explains in detail about the proposed architecture and methodology. Section 7.4 describes the experiments performed and evaluation metrices as well as the results in details. Section 7.5 concludes the work.

7.2 TECHNIQUES USED

7.2.1 Naïve Bayes

Naïve Bayes (NB) algorithm is a probabilistic classification algorithm based on Bayes' theorem. The assumption of Naïve Bayes includes independence of features, where features do not interact with each other while predicting the classification results [16]. The NB classifier is a supervised learning algorithm and can be efficiently trained using small amount of data. Since the variables or the features are assumed to be independent, thus only the calculation of the variances of each class variable is required instead of calculating the values of the entire covariance matrix. Naïve Bayes algorithm is remarkable in terms of performance because it can train on limited data as well [17–21]. Equation (7.1) describes the Bayes Rule below.

$$P(c_i|D) = \frac{P(c_i)P(D|c_i)}{P(D)} \tag{7.1}$$

where $P(C_i) = P(C = c_i) = \frac{N_i}{N}$

where $P(C_i)$ is the ratio of number of training samples of class C_i in dataset (i.e. N_i) over total number of samples in the dataset (i.e. N)

Although Naïve Bayes is based on the unrealistic assumption of independence between different features but still it delivers reasonable performance for many classification tasks. It is simple and elementary to implement. Therefore, it has been used by the researchers in different tasks [22, 23]. In [24], the Naïve Bayes classifier performed exceptionally well where the probabilities computed by the model were not important. It is used for different classification applications like spam filtering [25] in emails, spam categorization, and categorization of web content.

7.2.2 Long short-term memory model

Long short-term memory (LSTM) was proposed in [26] which has proven to be extremely powerful in the past for learning sequences over standard recurrent neural networks (RNNs). RNNs are basically neural networks unfolded over multiple time steps and thus can process sequential data such as time series and sentences. With each time step having hidden state, processing inputs from the previous hidden state and current inputs. The disadvantages of RNNs are vanishing and exploding gradients. The transition functions of the LSTM [26] are defined as per Equation (7.2)–(7.7):

$$i_t = \sigma\left(W_i \cdot \left[h_{t-1}, x_t\right] + b_i\right) \tag{7.2}$$

$$f_t = \sigma\left(W_f \cdot \left[h_{t-1}, x_t\right] + b_f\right) \tag{7.3}$$

$$o_t = \sigma\left(W_o \cdot \left[h_{t-1}, x_t\right] + b_o\right) \tag{7.4}$$

$$c_t = f_t \odot c_{t-1} + i_t \odot q_t \tag{7.5}$$

$$h_t = o_t \odot \tanh(c_t) \tag{7.6}$$

In the above equations, $\odot$ signifies the element wise multiplication operation. The notations are: current time step x_t, old hidden state h_{t-1}, the input

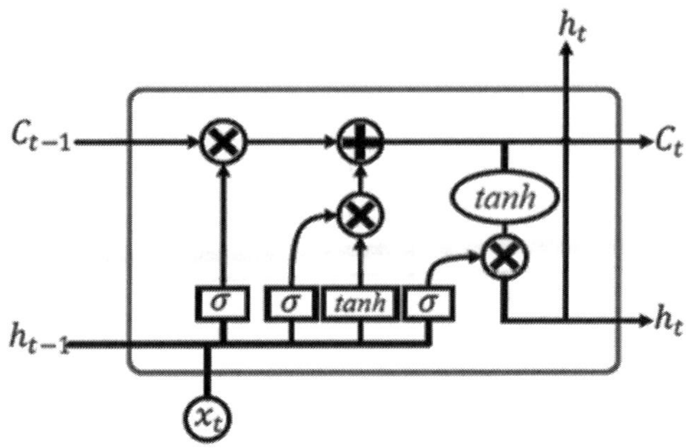

Figure 7.1 Architecture of LSTM cell.

gate i_t, the forget gate f_t, and the output gate o_t. Weights are denoted by W_o, W_q, and W_i which are trainable parameters with respect to different gates. Previous and current memory cell are denoted by c_{t-1} and c_t, respectively The current hidden state h_t is modified using c_t and is collectively decided by these gates. Figure 7.1 depicts the LSTM cell diagram.

7.2.3 Bidirectional long short-term memory (Bi-LSTM) model

Bidirectional LSTM (Bi-LSTM) models are helpful in extracting more contextual information [27]. The data processing is performed in the backward and forward direction simultaneously [28] and the LSTM outputs are merged using the sum statistic [29].

Bi-LSTM models serve as state-of-the-art models for classifying the sequential data into multiple classes and have been used extensively for speech recognition [30, 31, 32, 33, 34]. The text in the pages of the PDF documents is also a sequence of words and multiclass criteria also exist as the documents are to be classified into the respective language class. Therefore, Bi-LSTM model is applied for the classification task in this paper.

7.3 METHODOLOGY

7.3.1 Dataset description

Dataset comprises pdf files from three languages, viz. English, Slovakian, and Czech language. Data cleaning and preprocessing is performed over

Table 7.1 Dataset samples

Class	*Number of samples*
English	5000
Slovakian	5000
Czech	5000

the sentences. Count Vectorizer is used to encode set of sequences. Dataset is split as: 80% of data is used for training and remaining 20% is used in testing. Furthermore, cross validation is used to avoid the overfitting to a single sample set. Data are trained for 100 epochs which is selected by using grid search methodology. Table 7.1 shows the dataset details.

7.3.2 Feature extraction

Before starting with the analysis, firstly the extraction of the text from the PDF document file is performed. Tesseract OCR system is used to extract the content from a document page if the page is present in the form of a raster image [15]. After extracting, the text passes through document classification engine, explained in detail in Section 7.3.3.

7.3.3 Document classification engine

The document classification engine can assist humans in the separation of documents of different languages into different folders. This prevents the documents from getting lost if they are not properly labeled and accelerates the process of usage of these documents for proper application. The proposed classification engine can also ingest and classify the PDF documents containing the data in the form of images. Document classification engine serves as a prerequisite to classify documents to leverage extraction of information from a variety of PDF documents.

Figure 7.2 depicts methodological steps for carrying out classification. Initially, preprocessing is done, which includes removal of stop words using NLTK library in Python. Further missing values, interpolation, and stemming are performed in the dataset. Later on feature selection is performed.

Since most of the text was obtained from the image format pages of the PDF by applying the OCR on the raster scanned documents with noise, many out-of-vocabulary words were found in the extracted data. Such problem can be resolved either by running a named entity recognition (NER) [35] system or by the use of stemming and multiple regular expressions [36]. In this work, we have used stemming as a preprocessing step.

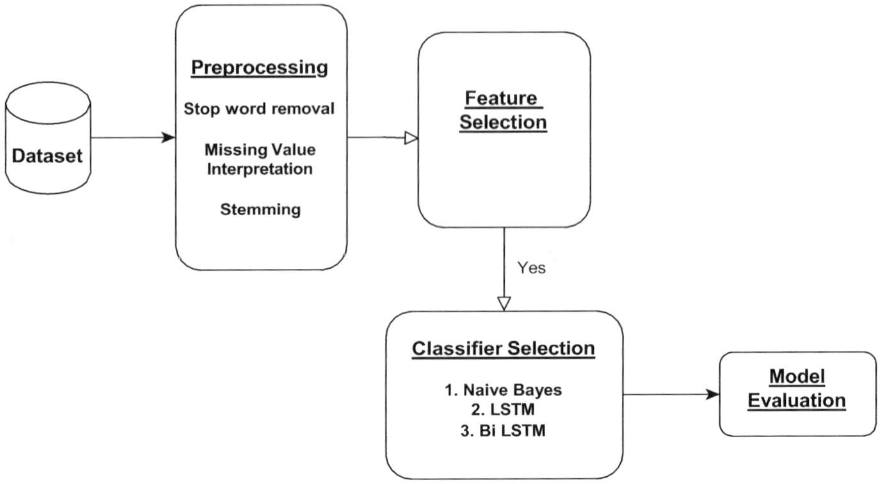

Figure 7.2 Methodological steps.

7.4 EXPERIMENTS AND RESULTS

7.4.1 Experimental details

For carrying out experiments, we have utilized Spyder 3.2.8 notebook with Python version 3.6.5. The experiments are executed on Windows 11, 64-bit OS system having 16GB RAM, with Intel(R) i5 CPU @ 2.40 GHz. Python Tensorflow and Keras API are used primarily to implement various algorithms in this work.

7.4.2 Baseline models

7.4.2.1 DNN: deep neural network

Architectural details of DNN are defined in Figure 7.3.

7.4.2.2 LSTM

Relu activation is used in all the layers except for last layer. Binary cross entropy loss is used as a loss function and for optimization, Adam is used for hyperparameter tuning across multiple epochs.

7.4.2.3 BiLSTM

The proposed architecture composed of embedding layer, spatial dropout layer, and a single Bi-LSTM Layer. We have used word embeddings of 128

```
Layer (type)                 Output Shape              Param #
=================================================================
dense_43 (Dense)             (None, 512)               1131008
_________________________________________________________________
dense_44 (Dense)             (None, 64)                32832
_________________________________________________________________
dropout_20 (Dropout)         (None, 64)                0
_________________________________________________________________
dense_45 (Dense)             (None, 32)                2080
_________________________________________________________________
dropout_21 (Dropout)         (None, 32)                0
_________________________________________________________________
dense_46 (Dense)             (None, 16)                528
_________________________________________________________________
dropout_22 (Dropout)         (None, 16)                0
_________________________________________________________________
dense_47 (Dense)             (None, 3)                 51
=================================================================
Total params: 1,166,499
Trainable params: 1,166,499
Non-trainable params: 0
_________________________________________________________________
```

Figure 7.3 DNN architecture.

dimensions as input layer for generating features of sparse input documents. To avoid overfitting, we have used spatial dropout of 1-dimensions with 40% dropping rate. We have further implemented Bi-LSTM layer along with further dropout rate of 20%. The learning parameters utilize grid search technique for optimization. Bi-LSTM layer utilizes 196 neurons as input. This choice of 196 neurons is selected after performing a grid search by selecting neuron range from 32 to 256 neuron sizes, and selecting the one that is giving good performance without overfitting. The final output layer utilizes Softmax function classifying multiple languages based on probabilities. Categorical cross entropy is used as a loss function. Adam is used for hyperparameter tuning as it has given good performance over other hyperparameter tuning algorithms. The output dense layer consists of three neurons for each of the classes. Figure 7.4 shows the architecture of the same.

7.4.3 Results

For evaluation of models, F-score and accuracy metrics are used in this work. Table 7.2 shows results for Bi-LSTM architecture.

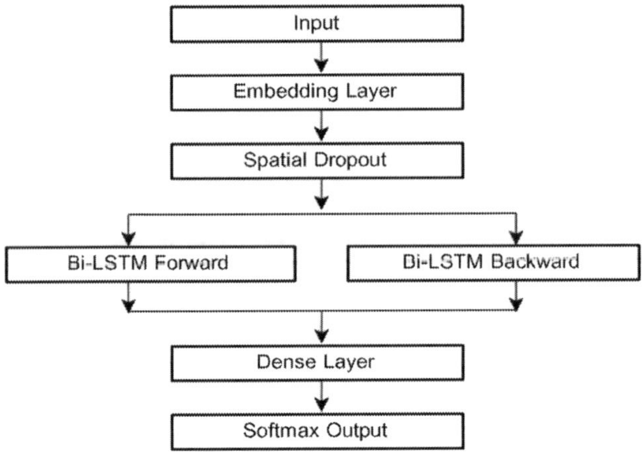

Figure 7.4 Model architecture using Bi-LSTM.

Table 7.2 Results for our proposed model

Evaluation metric	*Result*
F- score	0.98
Accuracy	98.73%

Table 7.3 Results for our proposed model

Model	*Accuracy*
Naïve Bayes	97.0%
DNN	97.1%
LSTM	98.2%
Proposed approach using Bi-LSTM	98.7%

Table 7.4 Comparative study with state-of-the-arts

Approach	*Model*	*Accuracy (%)*
Isa et al. [22]	Naïve Bayes	87.45
Van Linh et al. [12]	Multinomial NB	81
Van der Heijden et al. [1]	Meta Learning- L2 normalization	89.1
Isa et al. [22]	Naïve Bayes-SVM	88.02
Proposed approach	Bi-LSTM	**98.7**

Note: Bold value specifies the outperformance of the proposed approach over existing state-of-the-arts.

```
'Upozorňujeme, že jejím cílem je šetřit penězi za administrativní výdaje.\n',
'Zcela otevřeně přiznám, že ne vždy rozumím extrémně složitým vědeckým analýzám, které dostávám.\n',
'Při práci na této zprávě bylo třeba, abych pracovala právě s touto "představivostí".\n',
'A nakonec, plně podporuji myšlenku sjednotit vědeckovýzkumný sektor v energetické oblasti, jakožto klíčový prvek dané dlouhodobé politiky.\n'
'To je vážná věc.\n',
'Z tohoto hlediska zcela souhlasím s připomínkami pana Verheugena.\n',
'Rozprava je ukončena.\n',
```

(a)

```
'I can understand your approach a little bit.\n',
'Membership of Parliament: see Minutes\n',
'The establishment of a joint EU resettlement programme (\n',
'Decentralisation is very important for sustainable energy.\n',
'A majority rejected these solutions and wishes to stay in Bosnia.\n',
'(RO) At the start of the year, the unemployment rate reached 10%, while the deficit has grown in numerous Member States.\n',
'We are, as honourable Members know, supporting projects worldwide in the fight against racism, xenophobia or discrimination on any ground.\n'
'Agenda for next sitting: see Minutes\n',
```

(b)

```
'(DE) Vážená pani predsedajúca, papier môže byť všestranne užitočný, no je čoraz jasnejšie, že prestal byť dostatočne informatívny.\n',
'Ako viete, Európsky parlament sa chystá prijať uznesenie o dohode ACTA, pričom to však stále vyvoláva značné nepokoje a aj medzi našimi občanmi
to zďaleka nepanuje jednoznačný súhlasný názor. Myslím si preto, že by pre nám pomohlo, keby sme mali od vyjednávačov USA jasnú odpoveď ku dvom
bodom.\n',
'Mnohí z popredných vládnych predstaviteľov sú nezvestní.\n',
'Správa zdôrazňuje dôležitosť kolektívneho vyjednávania v boji proti diskriminácii žien predovšetkým v oblasti prístupu k zamestnaniu, platov,
pracovných podmienok, kariéry a profesijného vzdelávania.\n',
'Čína si zachovala svoje kultúrne tradície a hodnoty.\n',
'Musí sa preto využiť a zdokonaliť celý vývojový potenciál.\n',
```

(c)

Figure 7.5 Sample sentences extracted from Norwegian, Slovak, and English.

Table 7.3 shows a comparative analysis over multiple models. Results show the best performance of our proposed architecture using Bi-LSTM. Figure 7.5 shows sample classifications performed by the model.

Comparative study with state-of-the-art techniques is performed in Table 7.4. Table shows that the proposed method outperformed the existing ones with an accuracy of 98.7%.

7.5 CONCLUSION

Multilingual PDF document classification is an important problem that needs to be addressed owing to the ever increasing size of the electronic documents. This paper uses and compares different algorithms for the classification of multilingual documents. The Bi-LSTM algorithm produced accuracies consistently higher than other conventional algorithms like DNN and LSTM. The advantage of using Bi-LSTM is its ability to learn context in both the directions, which makes it more accurate than LSTM. Also Bi-LSTM requires very less computation requirements, hence it is used in this work. The proposed model has shown the best accuracy with an accuracy of 98.7%. In future, we can comparatively analyze results over much newer models keeping in mind low computing requirements.

REFERENCES

[1] Niels van der Heijden et al. Multilingual and Cross-lingual Document Classification: A Meta-Learning Approach. In *Proceedings of the 16th Conference of the European Chapter of the Association for Computational Linguistics*, pp. 196–197, 2021.

[2] C. D. Manning, P. Raghavan, H. Schütze et al. *Introduction to Information Retrieval*. Cambridge: Cambridge University Press, 2008.

[3] J. C. French, D. E. Brown, and N.-H. Kim. A classification approach to Boolean query reformulation. *JASIS*, 48(8):694–706, 1997.

[4] K. Kowsari, M. Yammahi, N. Bari, R. Vichr, F. Alsaby, and S. Y. Berkovich. Construction of FuzzyFind dictionary using golay coding transformation for searching applications. *International Journal of Advanced Computer Science and Applications*, vol. 6, pp. 81–87, 2015.

[5] T. Joachims. Transductive inference for text classification using support vector machines. In ICML, 99:200–209, 1999.

[6] S. Tong and D. Koller. Support vector machine active learning with applications to text classification. *Journal of Machine Learning Research*, vol. 2, no. Nov, pp. 45–66, 2001.

[7] M. Fernández-Delgado, E. Cernadas, S. Barro, and D. Amorim. Do We Need Hundreds of Classifiers to Solve Real World Classification Problems? *Journal of Machine Learning Research*, vol. 15, no. 1, pp. 3133–3181, 2014.

[8] E. Eskin, J. Weston, W. S. Noble, and C. S. Leslie. Mismatch String Kernels for SVM protein classification. In *Proceedings of the 16th Advances in Neural Information Processing Systems (NIPS)*, pp. 1441–1448, 2003.

[9] C. S. Leslie, E. Eskin, and W. S. Noble. The Spectrum Kernel: A string kernel for SVM protein classification. In *Proceedings of the 7th Pacific Symposium on Biocomputing*, vol. 7, pp. 566–575, 2002.

[10] A. McCallum, K. Nigam, et al. A comparison of event models for Naive Bayes text classification. In *Proceedings of the AAAI-98 Workshop on Learning for Text Categorization*, vol. 752, pp. 41–48, 1998.

[11] S.-B. Kim, K.-S. Han, H.-C. Rim, and S. H. Myaeng. Some effective techniques for Naive Bayes text classification. *IEEE Transactions on Knowledge and Data Engineering*, vol. 18, no. 11, pp. 1457–1466, 2006.

[12] N. Van Linh, N. K. Anh, K. Than, and C. N. Dang. An effective and interpretable method for document classification. *Knowledge and Information Systems*, vol. 50, no. 3, pp. 763–793, 2017.

[13] G. E. Hinton and R. R. Salakhutdinov. *Reducing the Dimensionality of Data with Neural Networks*. Science, vol. 313, no. 5786, pp. 504–507, 2006.

[14] R. Johnson and T. Zhang. Effective Use of Word Order for Text Categorization with Convolutional Neural Networks. In *Proceedings of the 2015 Conference of the North American Chapter of the Association for Computational Linguistics: Human Language Technologies*, pp. 103–112, 2015.

[15] R. Smith. An overview of the tesseract OCR engine. In *Proceedings of the Ninth International Conference on Document Analysis and Recognition (ICDAR)*, vol. 2, pp. 629–633. IEEE, 2007.

[16] H. Brücher, G. Knolmayer, and M.-A. Mittermayer. *Document Classification Methods for Organizing Explicit Knowledge*. Research Group Information Engineering, Institute of Information Systems, University of Bern, 2002.

[17] A. McCallum and K. Nigam. A comparison of event models for Naive Bayes text classification. *Journal of Machine Learning Research*, vol. 3, pp. 1265–1287, 2003.

[18] I. Rish. An empirical study of the Naïve Bayes Classifier. In *Proceedings of the IJCAI-01 Workshop on Empirical Methods in Artificial Intelligence*, vol. 3, no. 22, pp. 41–46, 2001.

[19] S. Ruan, B. Chen, K. Song, and H. Li. Weighted naïve Bayes text classification algorithm based on improved distance correlation coefficient. *Neural Computing and Applications*, vol. 34, nos. 4, pp. 2729–2738, 2022.

[20] P. Domingos and Michael Pazzani. On the optimality of the simple Bayesian classifier under zero-one loss. *Machine Learning*, vol. 29, nos. 2-3, pp. 103–130, 1997.

[21] S.-B. Kim, H.-C. Rim, D.-S. Yook, and H.-S. Lim. Effective methods for improving Naïve Bayes text classification. In *7th Pacific Rim International Conference on Artificial Intelligence*, vol. 7, pp. 414–423, 2002.

[22] D. Isa, L. H. Lee, V. P. Kallimani, and R. Rajkumar. Text documents preprocessing with the Bayes formula for classification using support

vector machine. *IEEE Transactions on Knowledge and Data Engineering*, vol. 20, no. 9, pp. 1264–1272, 2008.

[23] S.-B. Kim, K.-S. Han, H.-C. Rim, and S. H. Myaeng. Some effective techniques for Naive Bayes text classification. *IEEE Transactions on Knowledge and Data Engineering*, vol. 18, no. 11, pp. 1457–1466, November 2006.

[24] P. Domingos and M. J. Pazzani. On the optimality of the simple Bayesian classifier under zero-one loss. *Machine Learning*, vol. 29, nos. 2-3, pp. 103–130, 1997.

[25] T. S. Guzella and W. M. Caminhas. A review of machine learning approaches to spam filtering. *Expert Systems with Applications*, vol. 36, no. 7, pp. 10206–10222, 2009.

[26] J. Schmidhuber and S. Hochreiter. *Long short-term memory. Neural Computation*, vol. 9, no. 8, pp. 1735–1780, 1997.

[27] Z. Huang, W. Xu, and K. Yu. *Bidirectional LSTM-CRF Models for Sequence Tagging. arXiv preprint arXiv:1508*.01991, 2015.

[28] M. Schuster and K. K. Paliwal. Bidirectional recurrent neural networks. *IEEE Transactions on Signal Processing*, vol. 45, no. 11, pp. 2673–2681, 1997.

[29] A. Graves and J. Schmidhuber. Framewise phoneme classification with bidirectional LSTM and other neural network architectures. *Neural Networks*, vol. 18, no. 5, pp. 602–610, 2005.

[30] R. Ghaeini, S. A. Hasan, V. V. Datla, J. Liu, K. Lee, A. Qadir, Y. Ling, A. Prakash, X. Z. Fern, and O. Farri. DR-BiLSTM: Dependent Reading Bidirectional LSTM for Natural Language Inference. In *Proceedings of the 2018 Conference of the North American Chapter of the Association for Computational Linguistics: Human Language Technologies, Vol 1 (Long Papers)*, pp. 1460–1469, 2018.

[31] S. Hochreiter and J. Schmidhuber. Long short-term memory. *Neural Computation*, vol. 9, no. 8, pp. 1735–1780, 1997.

[32] E. Kiperwasser and Y. Goldberg. *Simple and Accurate Dependency Parsing Using Bidirectional LSTM Feature Representations. Transactions of the Association for Computational Linguistics*, vol. 4, pp. 313–327, 2016.

[33] A.Rao and N. Spasojevic. *Actionable and Political Text Classification Using Word Embeddings and LSTM*. Technical report, Lithium Technologies, July 2016.

[34] K. S. Tai, R. Socher, and C. D. Manning. Improved Semantic Representations from Tree-Structured Long Short-Term Memory Networks.In *Proceedings of the 53rd Annual Meeting of the Association for Computational Linguistics and the 7th International Joint Conference on Natural Language Processing*, vol. 1, pp. 1556–1566, 2015.

[35] N. Correia da Silva et al. Document Type Classification for Brazil's Supreme Court Using a Convolutional Neural Network. In *10th International*

conference on forensic computer science and cyber law (ICoFCS), Sao Paulo, Brazil, pp. 29–30, 2018.

[36] P. H. Luz de Araujo, T. E. de Campos, R. R. R. de Oliveira, M. Stauffer, S. Couto, and P. Bermejo. *Lener-br: A Dataset for Named Entity Recognition in Brazilian Legal Text.* In *Proceedings of the International Conference on the Computational Processing of Portuguese (PROPOR)*, Canela, RS, Brazil, vol. 13, pp. 313–323, September 24–26, 2023.

Chapter 8

FNDetector

Fake news detection using combinations of various features

Yun Hwa Wong (Jenny), Amanpreet Kaur, and Sangeeta Lal

8.1 INTRODUCTION

In the explosion era of technology and the modern world, fake news is permeating our daily lives at an alarming rate. The Cambridge dictionary has defined "fake news" as "false stories that appear to be news, spread on the internet or using other media, usually created to influence political views or as a joke" [1]. According to the world's leading public service broadcaster, British Broadcasting Corporation (BBC), the history of fake news started 2000 years ago during a civil war in the Roman Republic. Octavian wanted to win the war, and he disseminated the "fake news" against Mark Anthony by printing the slogan on the coins and distributing it to the public. In the 19th century, after the technique of print pressed was invented. One of the newspapers in 1835, articles were published about fantastic creatures such as flying batmen and unicorns living on the moon [2].

Before the concept of online social media came, most of the people get news via reading newspapers or listening to the broadcast channel on the radio. With the characteristic of connecting people and sharing content easily, the rate of disseminating news has increased drastically via online social media.

A study conducted by [3] on the Twitter dataset from the year 2006 to 2017 shows that between 1000 and 100000 people were affected by the top 1% of fake news while the truth was seldom spread to more than 1000 people. The study has shown how fast fake news can be disseminated on online social media.

Fake news is prevalent in many domains, especially in the domain of politics. During 2016 US presidential nomination process, the fake news about Donald Trump was greatly favored and has been shared extensively on social media. On Facebook, there were 115 fake news stories supporting Trump and 41 supporting Clinton [4]. The possible impact of fake news disseminated on public forum has drawn up particular attention from different researchers and the public. Fake news, however, is not just about political issues. In recent years, fake news can be disseminated widely

 DOI: 10.1201/9781003569619-8

Table 8.1 Example or descriptions of each approach to identify fake news

Approaches	*Examples/Descriptions*
Language approach	Bag of Words, Semantic Analysis, Deep Syntax
Topic-agnostic approach	Using the linguistic features and web mark-up capabilities
Machine learning approach	Deep learning neural network
Knowledge-based approach	Expert oriented, computational oriented and crowd sourcing oriented fact checking
Hybrid approach	Combination of different models or techniques

when a disaster or pandemic happens. The best example is the SARS-CoV-2 (COVID-19) virus breakout that happened in the year 2019. There are tons of fake news or false information related to information on how to get recover from the disease such as eating hot pepper can help to prevent or cure COVID-19, and drinking alcohol can protect the public against COVID-19 [5]. For example, fake news written as a Facebook post by a user regarding steaming can kill the Coronavirus and has been shared more than 33,000 times on Facebook [6]. Researchers have put great deal of time and effort to develop different algorithms and models to detect fake news. A study of categorizing different approaches used by various researchers has been done by de Beer and Matthee [7]. According to de Beer and Matthee [7], the approaches used by existing researchers can be categorized into five main categories (refer to Table 8.1). An example of each of the approaches can be found in Table 8.1.

Despite the myriad of methods and techniques proposed by researchers over the years to identify fake news, it remains the case that, as of now, no approach has achieved a level of detection with complete accuracy, to the best of our understanding. There are two main reasons:

Data challenge: Researchers can only use a few publicly available data sets, as far as we are aware, to develop and test their fake news detection models. Different data sets are suitable for different tasks. It is challenging to get the correct dataset such as the LIAR dataset introduced by Wang [8] that fulfills the research's requirements. However, these have limitations in terms of feature set availability and are not suitable for the work proposed in this paper.

Identification of an adequate learning model: The dataset and extracted feature have a significant influence on the learning model's performance [9]. Hence, it is important to carefully identify the features of a given data set. We have presented the FNDetector model. We used features from four domains: linguistic-based features, sentiment-based features, name entity recognition (NER)-based features, and textual features. We trained two variants of the FNDetector model, one using the Random Forest (RF) classifier and another using the long-term short memory (LSTM) classifier. We assessed the efficiency of the FNDetector model on Horne2017FakeNews Dataset. Experimental results show that FNDetector outperforms the baseline classifier significantly. The FNDetector model using the RF classifier outperforms the baseline classifier by 9.3% (F-measure) and 9.8% (accuracy) and FNDetector using the LSTM classifier outperformed the baseline classifier by 40.2% (F-measure) and 17.7% (accuracy). Following are the main research questions that we answered using this study.

RQ 1: How well the baseline classifier perform?
RQ 2: How well does the suggested model, FNDetector, perform in comparison to the baseline classifier?
RQ 3: How important to extract features from both title and contents of the article?
RQ 4: Which features are more important in the FNDetector model?
RQ 5: What is the consequence on FNDetector model's efficiency if the parameters of RF and LSTM classifiers are changed?

This is how the chapter is structured. We introduced fake news in Section 8.1. We will provide background information about fake news in Section 8.2. Section 8.3 reveals the particulars of contribution this study makes to the field of original research and will address related work. In Section 8.4, we will give detailed information about the proposed FNDetector model. In Sections 8.5, 8.6, and 8.7, we will provide details related to dataset extraction, experimental setup, and metrics for assessment. Section 8.8 will provide motivation, approach, and the results obtained for each Research Question (RQ). In Section 8.9, conclude and future directions will be provided.

8.2 BACKGROUND

Various forms of fake news can be circulated in our daily life. It is circulated in different formats such as videos, URLs, articles, etc. However, they are serving the same purpose and motivation which is the creator would like to get attention and unethical benefits from the public. Here, we will discuss different types of fake news that we encounter in our daily life and how the creator gets an advantage from it.

8.2.1 Clickbait

The clickbait can be represented by a link or thumbnail on online social media. It has become one of the dominant forms of medium to spread fake news on social media. The headline of clickbait is always misleading and can involve a curiosity gap from the public, gossip, or sensational aspect to attract the user to click on it [10]. A survey that has been conducted among 53 Standford students found that 96.2% of them came across clickbait stories online at least once each day [11]. The motivation for creating clickbait from the creator is to increase the number of websites visited and advertising avenues.

8.2.2 Propaganda

Propaganda is another form of disseminating fake news. Propaganda is usually used to distribute biased or misleading information that is related to political aspects or personal points of view before the election or protest happens in order to influence people's opinion. To achieve the greatest impact, propaganda employs strategies of subtle psychological and linguistic [12]. Due to the nature of propaganda, the heinous propaganda news sources have demonstrated their ability to have a significant influence. The power of propaganda and disinformation has been illustrated in recent events such as Brexit. Propaganda propagated on a notorious bus [13] has become a significant representation of the Brexit referendum campaign [14]. The false claim has been believed by half of the British public which has brought a negative impact during and after the referendum.

8.2.3 Satire and parody

The satire and parody is a humorous commentary on a work that calls for an imitation of the work or criticism of the world. The comic effect will be applied to both styles to entertain people. Although these do not usually classify as fake news, however, the information present in satire or parody might provide false information or mislead the public and bring a negative impact on the public's decisions. Examples of the website of satire and parody include The Onion (www.theonion.com/), The Daily Mash (www.thedailymash.co.uk/), etc.

In general, online social media and websites have become one of the dominant platforms for the public to get information. However, false information from social media or websites is difficult to distinguish from the public. Hence, the researchers have invested so much effort for identifying fake news.

8.3 LITERATURE SURVEY AND RESEARCH CONTRIBUTION

8.3.1 Literature survey

To train a model that supports high accuracy in detecting fake news, the dataset required for training the model has played an important role. It is challenging to find a reliable dataset to train a highly efficient model.

Shu et al., performed a study on extracted features from different datasets [15]. This extracted feature is separated into two different categories, i.e., news content and social context features. Linguistic-based and visual-based are feature representations types in news content features category. User-based, post-based, and network-based are categorized under social context features. The study discussed about extracted features from some popular datasets: BuzzFeedNews [16], LIAR [17], BS Detector [18], and CREDBANK [19]. The researchers have concluded that each of the datasets mentioned above has restricted some features from being extracted, for example: Only linguistic-based features can be extracted from the BuzzFeedNews dataset. The dataset is not supporting any existing feature extraction. Hence, to defeat the downside of existing dataset, the researchers have developed a new dataset, i.e., FakeNewsNet, to include all the news information and social context features.

Bhutani et al. [20] used three different datasets: LIAR [17], George McIntire7 [21], and merged dataset (the combination of the dataset from Kaggle [22], PolitiFact [23], and the Emergent dataset). They applied different text preprocessing techniques such as Bi-grams, Tri-grams, CountVectorizer, etc on the datasets. Additional columns such as tf-idf scores, sentiments, and cosine similarity scores have been added to the merged dataset. Therefore, by extracting the sentiment features from the false news detection model, accuracy of fake news identification is improved.

Khan et al. [24] and Iwendi et al. [25] carried out the investigation on the importance of feature extraction. Both of the studies extracted the feature from the text like the count of lowercase characters, average word length, positive sentiment score, etc., before passing them to advanced machine learning algorithms (RF algorithms, deep learning algorithms). The prediction precision of fake news detection before feature extraction by means of a decision tree classifier was 67.81%. However, after applying feature extraction, the result has significantly improved by 10% giving 77.58% prediction accuracy. Ajao et al. [26] included the sentiment element as one of the extracted features. The sentiment analysis used in the study was the emotional ratio score. The authors were using the latent semantic analysis and latent Dirichlet allocation to retrieve the emotional scores. As a result, the model which included emotional ratio gave better results than the existing extracted features that were used on the same dataset. Choudhary and Arora [9] developed three models to experiment on the prediction precision of fake news detection. The final result of the experiment shows Model 3 based on all linguistic features has achieved better accuracy over the other two models (Model 1 based on grammatical, sentimental, and syntax-based features and Model 2 based on readability features) which achieved 84.52% average accuracy.

The above studies have shown how sentiment analysis could improve the average accuracy of detecting fake news. Besides the traditional features such as textual and environment features that can be extracted to train model, Reis et al. [27] has introduced an extra new feature named as domain localization feature to improve model. The domain localization feature can include IP, city, country, latitude, and longitude extracted from the news website URL. By including the additional feature, the fake news detection model developed by the author manage to detect nearly all the fake news in their dataset.

Although the importance of feature extraction has brought up the attention of the researchers, however, the classification approaches used to build the fake news detection model are a force to be reckoned with. Khanam et al. [28] has performed a study on adopting different traditional classification algorithms for fake news. XGboost outperformed with an accuracy greater than 75%. Besides the traditional classification algorithm, the researchers such as [29] applied distinct the deep learning model to the dataset for comparison. As a result, most deep learning models gave superior performance than traditional classification algorithm. Though Naive Bayes is a commonly used method, it has great potential in detecting fake news, nearly matching the efficacy of deep learning models. On top of traditional and deep learning models, Khan et al. [30] has conducted a benchmark research on various machine learning models including the pretrained model such as BERT [31]. Result of pretrained BERT-based model was better than other models.

Furthermore, the researcher also tried to propose the latest technology such as blockchain as a solution [32]. There are a few challenges and limitations in adopting blockchain technology in fake news detection but the author believed that in the near future, detecting fake news would be much easier with the proposed strategy. Moreover, some existing libraries have been built, and the Flair library [33] is one of them. Flair allows the user to combine different kinds of word embedding such as BERT embedding, ELMo transformer embedding, etc [34]. Researcher in [35] has used the Flair library for building model. The study found out that by using the Flair library with a deep learning algorithm, the accuracy of detecting true and false information was up to 99.8%.

The study of distinguishing between fake news, real news, and satire was performed by Horne and Adali [36]. The researcher has used three different datasets including the Buzzfeed election data set, Burfoot and Baldwin data set, and random political news data set that was created by the researcher from different sources. Three main features: psychological, stylistic, and complexity features were extracted from the dataset and run on the SVM model with a linear kernel. Result shows that accuracy of separating fake and real news from body text of the article was 71% on the random political news dataset and 77% on the Buzzfeed election dataset.

Balshetwar et al. [37] worked on detecting fake news and analyzing sentimental behavior present in the text. They used sentiment present in the text to cues of emotional signals. They tested the proposed model on the ISOT and LIAR datasets. The projected approach achieved precision of 99.8%.

Uppada et al. [38] proposed a model that uses for text and images. They used a fine-tuned BERT model for text processing. Their proposed fusion model gave an accuracy of 93%.

8.3.2 Research contributions

We proposed the novel fake news detection model, i.e., FNDetector, using four different types of features (linguistic-based, sentiment-based, NER-based, and textual). We extracted these features from both title and content of the news. We present the results of an in-depth empirical analysis of the FNDetector model on a freely accessible dataset.

Algorithm 1 FNDetector Algorithm

```
Input: TD
Output: FNDetector
procedure FNDetector(TD)
ALGO = {Arf, ALST M}
for all A ∈ ALGO do
NTTemp = ExtractNewsTitleContent(TD)
NTCorpus = Preprocess(NTTemp)
F = ExtractFeatures(NTCorpus)
if A = Arf then
finalFeature = {FL, FS, FN, FT f}
if A = ALST M then
finalFeature = {FL, FS, FN, FT o}
FNDetector = learnClassifier(finalFeature, A)
return FNDetector
procedure Preprocess(TD)
return Lemmatization with POS(TD)
```

8.4 METHODOLOGY

Figure 8.1 presents the overview diagram, and algorithms 1 and 2 present the detailed steps of the proposed FNDetector model creation. For details about the notations utilized in the Algorithm 1 and Algorithm 2 refer to the Figure 8.2. FNDetector comprises two phases: model building and prediction. In phase 1, we train the model using the training data (step 1 to step 4). In stage 2, we use the FNDetector model to predict fake news (step 5). Following we describe all the steps shown in Figure 8.1 in detail:

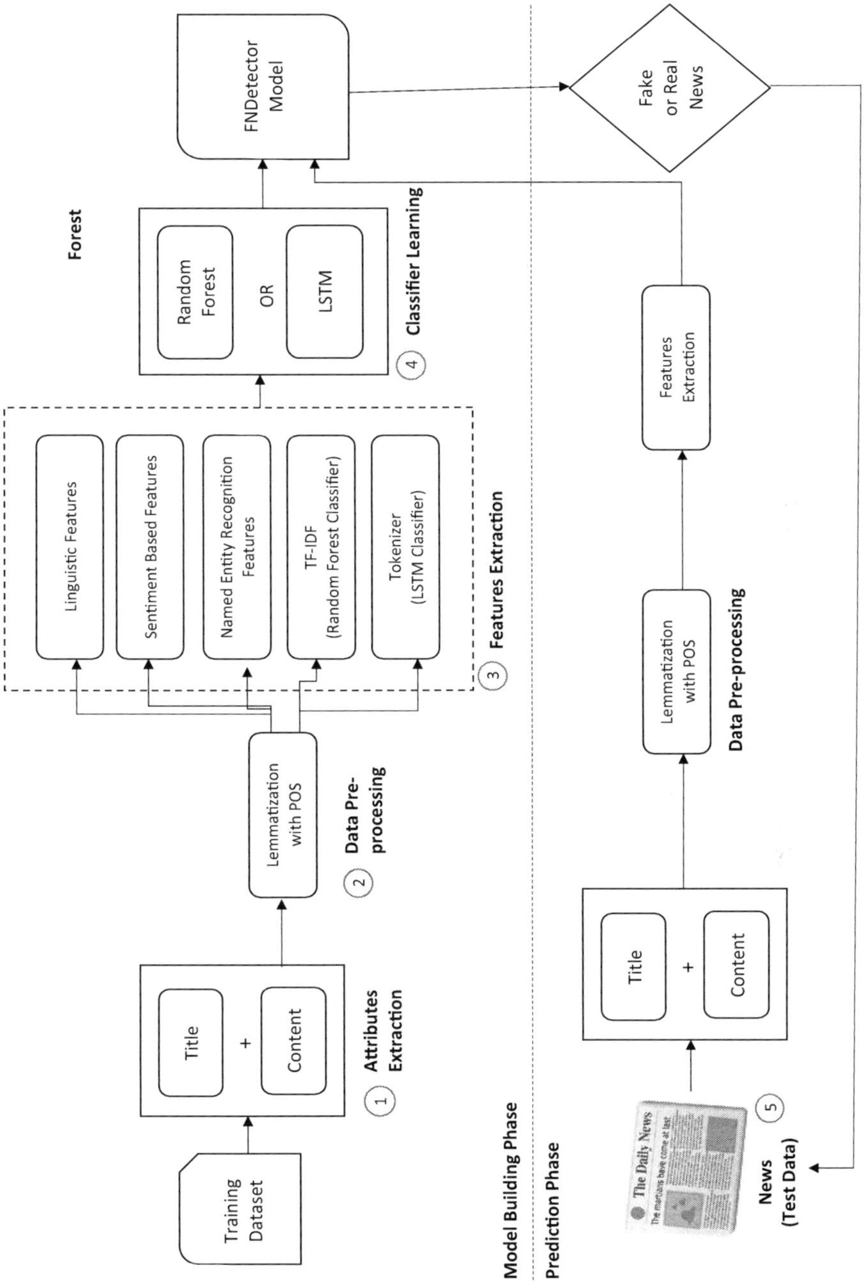

Figure 8.1 Overview diagram of the proposed FNDetector model.

Notation	Meaning	Notation	Meaning
NTCorpus	Lemmatized(with POS) title & content of news article	NT	One news article title & content
NT^T	News article title	NT^C	News article content
$\mathcal{F}_\mathcal{L}$	Lingustic-based features	$\mathcal{F}_\mathcal{S}$	Sentiment-based features
$\mathcal{F}_\mathcal{N}$	NER features	$\mathcal{F}_{Tf}$	Term Frequency – Inverse Document Frequency (Text pre-processing)
$\mathcal{F}_{To}$	Tokenizer (Text pre-processing)	$\mathcal{F}_\mathcal{L}^{iT}$	Linguistic-based feature for i^{th} news article from title of news
$\mathcal{F}_\mathcal{L}^{iC}$	Linguistic-based feature for i^{th} news article from content of news	$\mathcal{F}_\mathcal{S}^{iT}$	Sentiment-based feature for i^{th} news article from title of news
$\mathcal{F}_\mathcal{S}^{iC}$	Sentiment-based feature for i^{th} news article from content of news	$\mathcal{F}_\mathcal{N}^{iT}$	NER feature for i^{th} news article from title of news
$\mathcal{F}_\mathcal{N}^{iC}$	NER feature for i^{th} news article from content of news	$\mathcal{L}_{nn}^T$	Total number of nouns in title of news
$\mathcal{L}_{nn}^C$	Total number of nouns in content of news	$\mathcal{L}_{pro}^T$	Total number of pronouns in title of news
$\mathcal{L}_{pro}^C$	Total number of pronouns in content of news	$\mathcal{L}_{ad}^T$	Total number of adverbs in title of news
$\mathcal{L}_{ad}^C$	Total number of adverbs in content of news	$\mathcal{L}_{vb}^T$	Total number of verbs in title of news
$\mathcal{L}_{vb}^C$	Total number of verbs in content of news	$\mathcal{L}_{adj}^T$	Total number of adjectives in title of news
$\mathcal{L}_{adj}^C$	Total number of adjectives in content of news	$\mathcal{L}_{tw}^T$	Total number of words in title of news
$\mathcal{L}_{tw}^C$	Total number of words in content of news	$\mathcal{L}_{sw}^T$	Total number of stopwords in title of news
$\mathcal{L}_{sw}^C$	Total number of stopwords in content of news	$\mathcal{S}_n^T$	Total score of negative sentiment in title of news
$\mathcal{S}_n^C$	Total score of negative sentiment in content of news	$\mathcal{S}_p^T$	Total score of positive sentiment in title of news
$\mathcal{S}_p^C$	Total score of positive sentiment in content of news	$\mathcal{S}_{nu}^T$	Total score of neural sentiment in title of news
$\mathcal{S}_{nu}^C$	Total score of neural sentiment in content of news	$\mathcal{S}_c^T$	Compound score of sentiment in title of news
$\mathcal{S}_c^C$	Compound score of sentiment in content of news	$\mathcal{N}_p^T$	Total number of person entity in title of news
$\mathcal{N}_p^C$	Total number of person entity in content of news	$\mathcal{N}_o^T$	Total number of organization entity in title of news
$\mathcal{N}_o^C$	Total number of organization entity in content of news	$\mathcal{N}_g^T$	Total number of geopolitical entity in title of news
$\mathcal{N}_g^C$	Total number of geopolitical entity in content of news	$\mathcal{N}_m^T$	Total number of money entity in title of news
$\mathcal{N}_m^C$	Total number of money entity in content of news	i	Temp Variable
$\mathcal{A}_{rf}$	Random Forest algorithm	$\mathcal{A}_{LSTM}$	LSTM algorithm

Figure 8.2 Notations used in the FNDetector algorithm (1) and feature extraction (algorithm 2).

8.4.1 Training dataset collection

The experimental dataset is split into a test dataset and a training dataset. We use 80%–20% training and test split. We create models with a training dataset, and we evaluate model's performance with a test dataset. Prior to training any algorithm, the quality of the dataset must be guaranteed in order to develop a high-performance false news detection model. We load the dataset into a data frame with Pandas library. We shuffle the data before model creation to make sure the dataset is mixing up and not sticking to certain patterns according to the sorted order. We make sure that the dataset does not consist of any null values.

8.4.2 Attribute selection (step 1)

We selected the title and content of news articles for model creation (refer to step 5 in Algorithm 1). We selected both title and content of news articles because we hypothesize that both title and content consist of important information regarding the news and including both of these attributes can be beneficial for separating fake news from real news.

8.4.3 Preprocessing (step 2)

Lemmatization and stemming are popular preprocessing techniques. These techniques were utilized to enhance the performance of document retrieval precision [39]. According to Manning et al. [40], stemming is a rudimentary heuristic process that removes the end of the word and derivational affixes from the word to achieve the goal. Compared to stemming, lemmatization is the process of employing vocabulary and morphological analysis to eliminate a word's inflectional endings and return the word to its base or dictionary form. Table 8.2 shows the example of stemming and lemmatization.

In our work, we apply lemmatization to both title and content of the news articles. We did not apply stemming techniques in our dataset as lemmatization achieves improved accuracy and result in comparison to stemming which the result has been shown in the study [39]. However, the default behavior of lemmatization that changes the word's base to a noun might bring the possibility of reducing the accuracy of our final result outcome. Hence, to minimize the mistake that was brought by default lemmatization, we combine the part-of-speech (POS) tagging and lemmatization in preprocessing the news title and content (refer to step 6 of Algorithm 1). The POS tag, adjective, noun, verb, and adjective are used to identify the specific text in the dataset. So, when the lemmatizer detected a word in the verb from the text, instead of changing it to a noun by default, it will return the base format of the verb. In this work, we are using the lemmatizer from Wordnet [41]. Wordnet is a huge and publicly accessible English lexical database that is used to create organized semantic associations between words. Table 8.3 shows some of the examples from the dataset after lemmatization.

Table 8.2 Example of stemming and lemmatization

Type	*Before processed*	*After processed*
Stemming	Studies, Wander News	Studies, Wand New
Lemmatization	Studies. Running breathed	Study, run breath

Table 8.3 Example of lemmatization from dataset

Before lemmatization	*After lemmatization*
German prosecutors have closed an investigation into suspected mass phone tapping of German citizens by British and U.S. spies after finding no concrete indication of any criminal activity, they said on Thursday	German prosecutors have closed an investigation into suspected mass phone tapping of German citizen by British and U.S. spy after finding no concrete indication of any criminal activity, they said on Thursday

8.4.4 Feature extraction (step 3)

As we mentioned in Section 8.3, feature extraction is crucial to create a high-accuracy model. We extracted features from four different domains. We believe that features extracted from different domains can be beneficial in capturing different aspects of the news, and hence, can aid in improving accuracy (refer to step 7 of Algorithm 1). We extracted 32 features to train the FNDetector model.

8.4.5 Linguistic-based feature

The linguistic-based features are used to categorize the word or phoneme specified in particular pillars such as syntax, semantics, morphology etc. According to Mahyoob et al. [42], words play a significant part in news and the style of writing the news's contents brings a huge impact on the public. Conroy et al. [43] has mentioned that even though the creator of fake news tries to keep their speech under control, there will still have some verbal "leakage" that is difficult to be controlled such as the pronoun frequencies, use of verbs, conjunction patterns etc. Hence, to detect the "leakage" or "predictive deception indicators" in message content, we need to extract the linguistic-based feature from the dataset. There are two linguistic features we will be focusing on which are grammatical feature and lexical feature:

Grammatical-based feature refers to the structure of the sentences which consists of syntax, morphology, phonology, and semantics. In this project, we extracted the count of noun, verb, adjective, pronoun, and adverb from news headline and body from the dataset. To extract the grammatical feature, we used the concept of POS tagging. POS tagging is also known as grammatical tagging which each of the word sequences in a sentence will be automatic assign or tagged with a grammatical classification which normally contains a verb, adjective, pronoun, noun, etc [44]. Tagging each of the words with different grammar features in a large dataset is tedious and time-consuming. Hence, we took the help of the Natural Language Toolkit (NTLK) to perform the POS tagging.

Lexical-based feature is another type of linguistic-based feature. Lexical is referring the relationship between a word or vocabulary of a language. In our project, we focused on the lexical frequency where the frequencies of the text with correspondences occur in a large corpus such as the number of stopwords like "the", "is", etc appear in the title or contents of the article. Besides, we also calculated the count of words in each article.

Table 8.4 provide the details of all the 14 linguistic-based features extracted in this study.

8.4.6 Sentiment-based feature

Sentiment analysis is found to be useful in identifying suspicious accounts from online social media by extracting the element of feelings, attitudes, and opinions [45]. It is also found useful in fake news detection when combined with other extracted features [26]. This happens because fake news often is biased toward one type of emotion. When people are experiencing positive or neutral emotions, they are more inclined to believe a claim to be "true," yet when they are experiencing negative emotions, they may become more skeptical [46].

Sentiment polarity is one of the important elements in sentiment analysis. It helps to determine how the stated sentiment is oriented. For example, it can help to identify if the text is expressed in negative, neural sentiment, or negative from a statement. In this project, we extracted the polarity scores as one of the features using the sentiment analyzer from the natural language

Table 8.4 Details of the Linguistic-based features extracted

Feature's name	*Data type*	*Feature domain*	*Description*
noun content	N	Grammatical-based	Count of nouns in content
noun title	N	Grammatical-based	Count of nouns in title
pronoun content	N	Grammatical-based	Count of pronouns in content
pronoun title	N	Grammatical-based	Count of pronouns in title
adverb content	N	Grammatical-based	Count of adverbs in content
adverb title	N	Grammatical-based	Count of adverbs in title
verb content	N	Grammatical-based	Count of verbs in content
verb title	N	Grammatical-based	Count of verbs in title
adj content	N	Grammatical-based	Count of adjectives in content
adj title	N	Grammatical-based	Count of adjectives in title
totalWord content	N	Lexical-based	Count of words in content
totalWord title	N	Lexical-based	Count of words in title
totalStopWords content	N	Lexical-based	Count of stopwords in content
totalStopWords title	N	Lexical-based	Count of stopwords in title

Note: N, numeric.

Table 8.5 Details of the sentiment-based features extracted

Feature's Name	*Data type*	*Description*
neg content	N	Count of negative sentiment in content. Neg
title	N	Count of negative sentiment in title. Pos
content	N	Count of positive sentiment in content. Pos
title	N	Count of positive sentiment in title. Neu
content	N	Count of neural sentiment in content. Neu
title	N	Count of neural sentiment in title
compound content	N	The compound score of sentiment in content
compound title	N	The compound score of sentiment in title

Note: N, numeric.

toolkit (NLTK) package [47]. There are four different scores in this package which are positive, negative, neural, and compound score. The polarity score of each score can be ranged from −1 to +1 and the add-up of positive, negative, and neural scores will equal 1. The compound score is total of positive, negative, and neural scores with the range of most extreme negative (−1) to most extreme positive (+1). The more positive the text is, the closer the compound score is to +1. We extract 8 sentiment-based features. Table 8.5 shows the details of the sentiment-based features.

8.4.7 Named entity recognition (NER)-based features

NER is a procedure used to identify and classify the designated entities that are presented in a document. According to Shelar et al. [48], the NER tool usually scans through the whole text document and categorizes the entity such as place, organizations, date, major people and etc. It allows for easy content discovery by categorizing the articles in defined hierarchies with the known relevant tags for each text document. With the help of the NER approach, we can identify the relationship between the entities and the truth of news or articles. For example, when an election happens, fake news is tended to have major people's names as headlines to attract the public. In this work, we are using the trained model "en core web sm" (an English pipeline that includes vocabulary, syntax, and entities with training on written web text documents such as comments, news, and blogs) from the Sparcy library. Table 8.6 shows the example of entities of the NER approach in the trained model. A study from [3] has shown that most of the fake news is related to political news as the false information or political rumors on Twitter spiked during the 2012 and 2016 US presidential elections. This has motivated us to focus on the entities that are more relevant to political news such as organization (ORG), the major person (PERSON), geopolitical (GPE), and money (MONEY) while developing the fake news detection model. We extract eight NER features. Table 8.7 shows the details of all the NER-based features extracted.

Table 8.6 Example of named entities in Sparcy library

Entities	*Examples*
ORGANISATION	Apple, Tesla, Amazon
PERSON	Donald Trump, Elon Musk
MONEY	1 billion dollar, 2 pound
GPE	India, Germany
LOCATION	Mount Titlis, Yellow River
DATE	5 JUL, 24 January 2022
TIME	two-forty am, 11.09 P.M

Table 8.7 Details of the NER-based features extracted

Feature's Name	*Data type*	*Feature Domain*	*Description*
person content	N	NER	Count of person in content
content. person title	N	NER	Count of person in title
org content	N	NER	Count of organisation in content. org
content. org title	N	NER	Count of organisation in title. gpe
title. gpe content	N	NER	Count of geopolitical in content. gpe
content. gpe title	N	NER	Count of geopolitical in title. money
money content	N	NER	Count of money in content. money
money title	N	NER	Count of money in title

Note: N, numeric.

Table 8.8 Details of the textual features extracted

Feature's Name	*Data type*	*Feature Domain*	*Description*
title_lem	T	Textual	The lemmatized (with POS) title of articles and news
content_lem	T	Textual	The lemmatized (with POS) content of articles and news

Note: T, textual.

8.4.8 Textual features

We have also used textual features in the FNDetector model. We use lemmatized title (with POS) and lemmatized content (with POS) (refer to Table 8.8). However, because textual features cannot be directly fed into the classifiers. We used Term Frequency – Inverse Document Frequency (TF-IDF) (RF classifier) and tokenizer (LSTM classifier). Following are more details about TF-IDF and Tokenizer. TF-IDF: According to Qaiser and Ali [49], the term frequency (TF) is referring the occurrence of a particular term in a document. To calculate the inverse document frequency of a particular term, we can use the following formula:

$$\text{TF} = \frac{\text{Occurrence of Words}}{\text{Count of words of document}} \tag{8.1}$$

For example: if a 2000 words document contains the word "library" 5 times. The TF of the word "library" will be 5/2000, which is 0.0025. All the terms play an important role in a document. Inverse document frequency (IDF) aims to assign a weight to a particular term. The IDF algorithm assigns a more weightage to an infrequent term and a lower weightage to a frequent term in the document. To calculate the IDF of a particular term, we can use the following formula:

$$\text{Inverse Document Frequency}(\text{IDF}) = \log_e \frac{\text{Total number of document}}{\text{Number of documents with term}} \tag{8.2}$$

For example, if we have 20 documents and the term "library" appears in three documents, the IDF of the term will be loge(20/3) = 1.897. In order to address issues like stopwords that are highly prevalent in documents but have limited significance for text mining, the researchers have suggested merging TD and IDF computations to represent a term's relative importance to a document within the collection. The Term Frequency – Index Document Frequency (TD-IDF) will assign higher importance to the term that has higher frequency but less occurrence in the total number of documents. To calculate TD-IDF, it is just simply taken the multiplication of TD and IDF. For example, to calculate the term "library" in the collection, we will multiply 0.0025 and 1.897 which is 0.00474.

The concept of TD-IDF is playing important role in training the fake news detection model. By using the concept of TD-IDF, we can analyze what are the common terms being used in fake news to attract the public. In our work, the TD-IDF will be applied to the lemmatized title and the content of the articles. To implement the TD-IDF feature in the dataset, the TfidfVectorizer in the scikit-learn package [50] (a machine learning library) is being used.

Tokenizer: The TfidfVectorizer from the scikit-learn package will return the result as a sparse matrix. However, LSTM classifier gives better performance when the sequence is used as input. Hence, for the LSTM classifier, we vectorized heading and body of news article by turning each text into a sequence of integers using Tokenizer [51] (from the TensorFlow package).

8.4.9 FNDetector classifier learning (Step 4)

We combine all the features, i.e., linguistics-based, sentiment-based, NER-based, and textual features. Using this, we obtain a total of 32 features. Using these we train two versions of the FNDetector model, one using the RF classifier ($\text{FNDetector}_{\text{RF}}$) and the other using the LSTM classifier

Table 8.9 The configuration details of RF and LSTM classifiers used in this work

Method	*Setting*
RF Classifier	n estimators = 100
LSTM	epochs = 10

($FNDetector_{LSTM}$). The configuration related details of both the algorithms are shown in Table 8.9. Following are the details of both of the algorithms: Random Forest (RF) Classifier: Breiman et al. [52] first presented the RF method, an ensemble learning approach for regression trees and classification. A collection of classifiers with tree structures make up the RF classifier, and each tree votes for the class that is most popular.

Compared to other machine learning algorithms, the random forest doesn't consist of overfitting problems due to the use of the Strong Law of Large Numbers. The core method for the random forest to improve its accuracy was using the bagging method (also known as Bootstrap aggregating [53]. Besides, the out-of-bag method was used to estimate the strength, correlation, and generalization error. Furthermore, based on the strength of each predictor and their correlations, the algorithm offers information about the random forest's prediction ability.

According to Breiman [52], the random forest provides significant improvement in classification from the concept of expanding the collection of trees and letting them choose the most popular class.

Figure 8.3 describes RF algorithm. Taking into account that the training dataset includes balls of various hues, including pink, green, blue, yellow, and red.

The dataset will be divided into subsets and assigned to one decision tree in the RF algorithm. Each of the decision trees produces one output. For example, decision tree 1 and 2 produces yellow ball as their output, and tree n produces blue ball as their output. Then, the random forest algorithm will collect the majority voting and produce the final prediction. In our example, most of the tree has selected the yellow ball as their prediction output. Hence, the final prediction of RF will be the yellow ball.

Many studies such as [54] have shown that RF is among the machine learning methods with the highest performance, particularly for prediction tasks. The RF algorithm was formed by a tree-based model. The random forest algorithm is being used widely across different areas such as stock price prediction in the financial industry [55]. There are also lot of researchers who had used the random forest as the classifier to train fake news detection model as mentioned in Section 8.3. In our work, we tend to use different extracted features with RF algorithms to experiment with prediction accuracy.

Long-short term memory neural network (LSTM): The pioneer of the LSTM is the recurrent neural network (RNN) that utilizes consecutive or

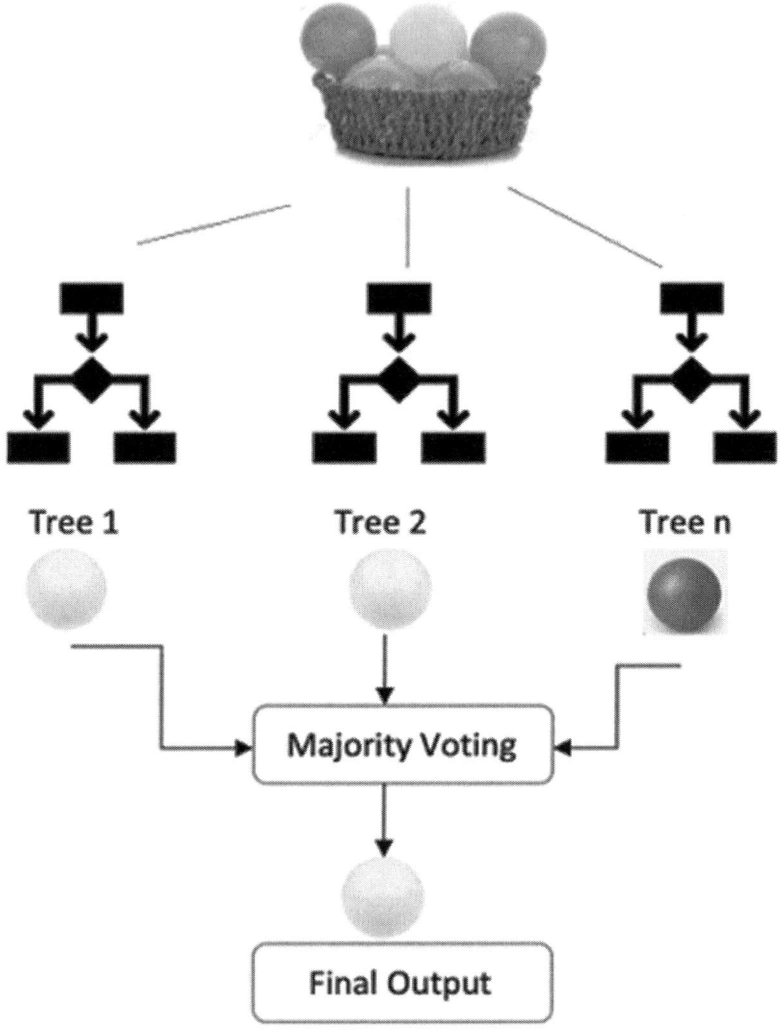

Figure 8.3 An illustrative example of RF algorithm.

time series data. The RNN is often used in time series forecasting such as forecasting the next twelve-month energetic usage of a household. According to Hochreiter and Schmidhuber [56], the existing RNN causes the vanishing and exploding of gradients when the error signals are flowing backward. To overcome this issue, Hochreiter and Schmidhuber [56] has introduced the concept of LSTM. The LSTM was formed by four core elements which are forget gate, input gate, output gate, and a complex unit called memory cell

[57]. Different from other traditional deep neural networks, the memory cell in the LSTM algorithm has the center linear unit with fixed self-connection. It goes along the whole chain and makes the selection on which information needs to be stored with the help of three gates [56]. This process allowed the LSTM network to identify long-term dependencies and memorize multiple time dependencies with different characteristics [58].

In our work, we have created the LSTM neural network to train the model. Two embedding layers are created to store the information of the title and content of the article. Then, two LSTM layers with 128 neurons are created to hold the output of embedding layers. A concatenate layer is created to concatenate the output of the two LSTM layers. Furthermore, a dense layer (a layer of neurons that receive the input from the previous layer) is created to process the other numeric features. Another concatenate layer is created to process outcomes from previous two layers before going into another dense layer. Using the input from former layer, the final output layer produce one result as the end product.

The overall LSTM neural network model for our project can be viewed in Figure 8.4. We used the sigmoid function as an activation method for our final output layer. This helps network to learn which data needs to be forgotten or remembered by holding the value between 0 and 1. The information will be forgotten if the value is 0 and stays if the value is 1. Since our final result has only two outcomes which are true or false news, we used binary cross entropy function for providing better performance for the binary classification model.

To perform the experiment using LSTM and answer the research questions in Section 8.8, the maximum number of most frequent words have been defined to 3000 words and the maximum number of words in each title or contents of the article to 300 words. Then, the title and contents of the article will be tokenized by Tokenizer. Three different inputs with the shape of 300 (for the title of the article), 300 (for the contents of the article), and 30 (the other 30 numeric features) will be created for model training. For all the LSTM classifiers, we have trained them with 10 epochs.

To make sure the training model in all experiments presented in Section 8.8 are consistent, we have defined the basic setting for the random forest and LSTM neural network classifier which is shown in the following table:

8.4.10 Prediction phase (step 5)

In this phase, we applied the FNDetector model to guess the label of news articles available in the test dataset. In order to accomplish this, we extract all 32 characteristics after extracting the news items' titles and contents from the test dataset. To determine whether or not the news article is phony, we feed these features into the FNDetector model.

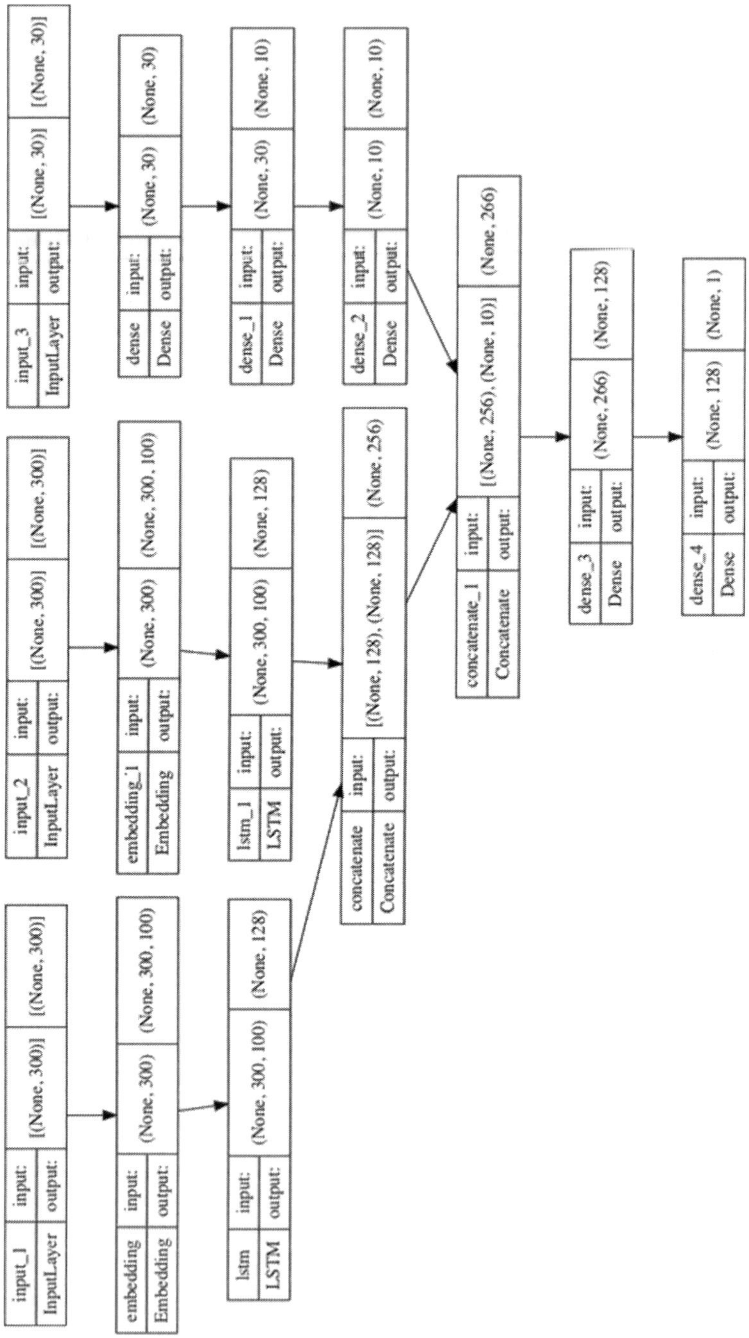

Figure 8.4 The LSTM neural network.

Table 8.10 Summary of the dataset

Dataset name	*News categories*	*Titles count*	*Contents count*
Buzzfeed political news	Fake	48	48
	Real	53	53
Random political news	Fake	75	75
	Real	75	75

8.5 DATASET EXTRACTION

In this work, we use Horne201 FakeNews Data [36]. The particulars about given dataset are shown in Table 8.10. There are two popular news datasets: Buzzfeed news and random political news. The first Buzzfeed news dataset contains 48 fake news and 53 real news, whereas the random political news dataset contains 75 fake news and 75 real news, respectively. These datasets contain two pieces of information which are the title and content of news. We downloaded the dataset from the repository https://github.com/rpitrust/fakenewsdata1.

8.6 EXPERIMENTAL SETUP

We use Python [59] for programming. The libraries and their associated process applied for our work are mentioned below:

- Framework: The algorithm was modeled using Jupyter Notebook for Python 3.
- Preprocessing: wordnet from nltk library, tf-idf vectorizer from sklearn library, pandas for data frame.
- Machine learning models: Random Forest classifier from sklearn library.
- Deep learning model: LSTM neural network from Keras.
- Features extraction: sklearn, nltk, spacy library.
- NLP: nltk library.
- Visualization: matplotlib, seaborn libraries.

8.7 EVALUATION METRICS

To make sure the classification or prediction model is behaving correctly and achieving the goal, a set of checklists will be applied to it to perform the measurement which is known as evaluation metrics. The performance and efficacy of classifiers can be assessed using evaluation metrics [60]. Several factors need to be considered when choosing the correct evaluation metric such as less complexity and less computational cost, the issue

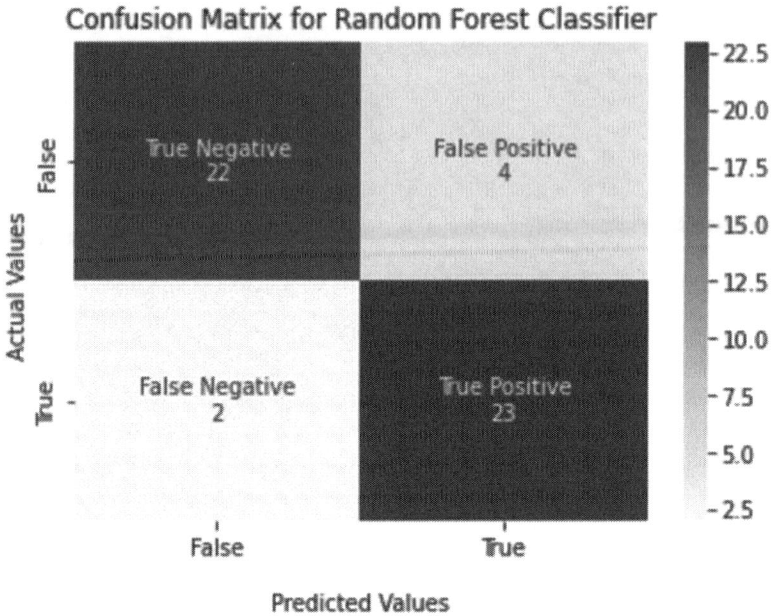

Figure 8.5 The example of confusion metric of Random Forest classifier.

of multiclass problems, informativeness, distinctiveness, and discriminable [61]. To ensure a classification model has high efficiency and high performance, it needs to make sure the predicted outcome is the same as the true outcome. The problem discussed in this paper is a binary classification problem. The evaluation metric applied for binary classification problems is the confusion metric which also belongs to the threshold types of evaluation metrics [61].

8.7.1 Confusion metric

The confusion metric is presented in a tabular format. It contains information about the actual and predicted class that is done by a classification model. There are four types of indicators in the confusion metric which are True Positive (TP), True Negative (TN), False Positive (FP), and False Negative (FN) [61].

The confusion metric provides the visualization to the researchers on the indicators so they can take further action on improving the classification model if necessary. Figure 8.5 shows how a confusion metric of random forest classifier. The x-axis and y-axis signify predicted values and real values, respectively. The darker the color in the box of the confusion metric, the higher the values.

8.7.2 Accuracy

The range of accuracy scores is between 0 and 1. In general, the accuracy score calculated the percentage of exact predictions over the total number of samples that are assessed [61]. The equation is listed below:

$$\text{Accuracy} = \frac{\text{TP} + \text{TN}}{\text{TP} + \text{TN} + \text{FP} + \text{FN}} \tag{8.3}$$

8.7.3 Precision, Recall, and F-Measure

Accuracy score is not a great evaluation metric to judge the performance of a classification model, as it provides a less distinctive value [61]. The accuracy score might not be accurate when the dataset is imbalanced. To overcome the limitation of the accuracy score, we can calculate the precision score of the classification model. The concept of calculating a precision score is using the sample that is precisely predicted from the entire predicted samples in a positive class [61]. Different from precision, the recall score can also be used to calculate the number of positive classes that are correctly classified out of all positive examples in the dataset. The harmonic mean of precision and recall is the f-measure (F1 score). Joshi [62] has mentioned that although the f-measure is less popular being used as the performance metric; however, it performs better than the accuracy score when come to optimizing the classifier for binary classification problems. The range of f-measure can be between 0 and 1. The formulas for computing these measures are listed below:

$$\text{Precision}(p) = \frac{\text{TP}}{\text{TP} + \text{FP}} \tag{8.4}$$

$$\text{Recall}(r) = \frac{\text{TP}}{\text{TP} + \text{TN}} \tag{8.5}$$

$$F \text{ measure} = \frac{2^* p^* r}{p + r} \tag{8.6}$$

8.7.4 Receiver operating characteristic (ROC) curve and area under ROC (AUC)

The receiver operating characteristic (ROC) has been widely used as a performance measurement for evaluating medical diagnostic systems [63]. According to Park et al. [64], the ROC is represented as a graph that consists of two key components which are sensitivity (or true positive rate) and specificity (or false positive rate), and it can differentiate the signal from the noise. In a ROC graph, the sensitivity will be plotted on the y-axis and specificity on the x-axis, respectively. The ROC graph can clearly show the relationship between the false positive rate and the true positive rate. The extension of the ROC curve, area under the ROC curve (AUC) has been proved as a useful metric for summarizing the efficacy of the test [65]. Mandrekar [65] also mentioned that the value of AUC can be ranged from 0 to 1, where 0 indicates the test was perfectly inaccurate and 1 indicates that test is perfectly accurate. In general, when the AUC is greater than 0.5, there is a high chance that the classifier can differentiate the positive class value from the negative class value. However, there are some cases where the AUC value is equal to 0.5, and the ROC curve is presented as the diagonal line on the graph. In this case, the classifier will not have the ability to differentiate between the positive and negative classes.

8.8 EXPERIMENTAL RESULTS

Here, we provide the details of experimental results.

RQ1: What is the performance of the baseline classifier?

Motivation: We worked on building a fake news detection model using different types of features. It is significant to recognize how the fake news detection model performs without using any of the external feature extraction. Hence, in this RQ, we built our baseline classifier using the title and the content of news articles.

Approach: We used the textual data, i.e., the lemmatized title and content of the news articles to build the baseline classifiers. We then vectorize the dataset by using TfidfVectorizer before training the RF classifier model. For the LSTM classifier, we tokenize the textual data and insert it into an embedding layer before going to the LSTM layer. For this experiment, we use 100 n estimators for RF and 10 epochs for LSTM.

Results: Figure 8.6 and Table 8.11 show the results of this RQ. We obtained an accuracy value of 60.7% and 78.4% for RF and LSTM classifiers, respectively, and an F-measure of 79.2% and 41.1% for RF and LSTM classifiers, respectively (refer to Table 8.10). These results indicate that the baseline classifier gives very low accuracy and F-measure with respect to fake news detection, and hence, there is a need to improve these models. The baseline classifiers perform poorly for fake news detection tasks with RF and LSTM classifiers giving an accuracy of 60.7% and 78.4%, respectively. The performance of the LSTM classifier is much lower than the RF classifier both in terms of accuracy and F-measure.

RQ 2: What is the performance of the proposed model, i.e., FNDetector, when contrasted to the baseline classifier?

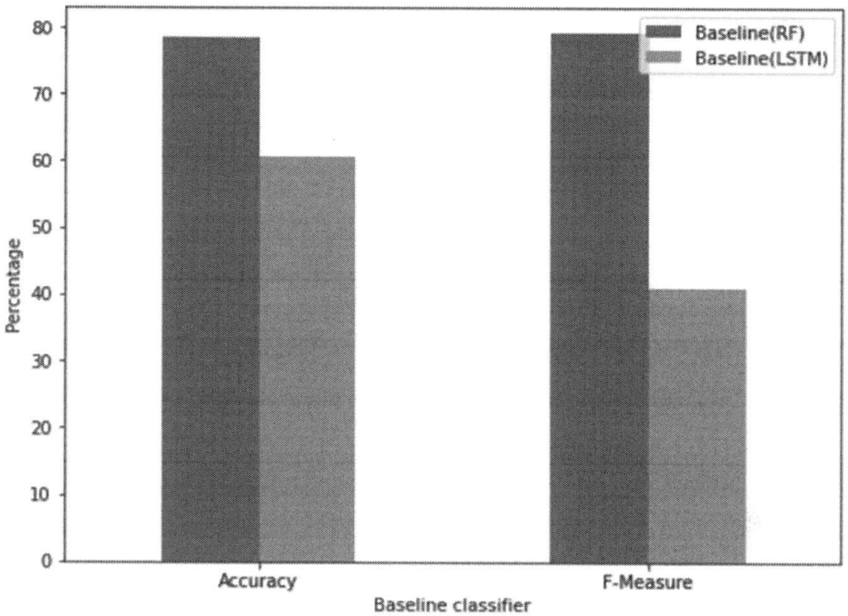

Figure 8.6 The performance of the baseline classifier.

Table 8.11 The performance of baseline classifiers

Method	*Accuracy (%)*	*Precision (%)*	*Recall (%)*	*F-Measure (%)*
Baseline_{RF}	78.4	80.8	77.8	79.2
Baseline_{LSTM}	60.7	100	25.9	41.1

Motivation: In this RQ, we investigated the performance of the proposed FNDetector model. To build this model, we combined different types of features. Additionally, we extracted features from both the title and content of news articles. This RQ will help in providing insights into the effectiveness of the FNDetector model.

Approach: To answer this RQ, we trained the FNDetector model using RF and LSTM classifiers to build two variants of the FNDetector model, i.e., FNDetector_{RF} and FNDetector_{LSTM}. We tested the performance of these two models on the test dataset and computed various metrics. We also performed a comparison of these two models with the baseline classifiers.

Result: Table 8.12 and Figure 8.7 show the prediction results of the FNDetector_{RF} and FNDetector_{LSTM} models. Table 8.12 shows that

Table 8.12 The performance of the FNDetector model

Method	*Accuracy*	*Precision*	*Recall*	*F-Measure*
FNDetector_{RF}	88.2%	85.2%	92.0%	88.5%
FNDetector_{LSTM}	78.4%	74.0%	88.9%	81.3%

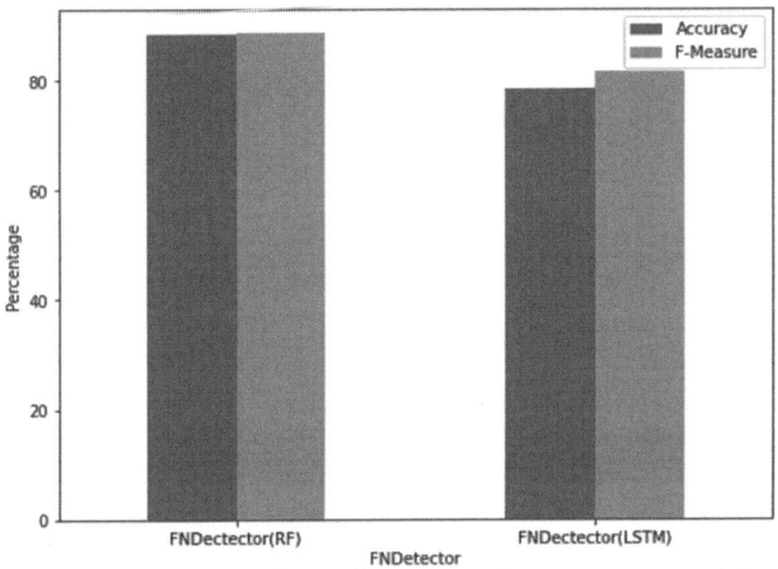

Figure 8.7 The performance of FNDetector model.

FNDetector_{RF} gives an accuracy of 88.2% and an F-measure of 88.5%, whereas FNDetector_{LSTM} gives an accuracy of 78.4% and an F-measure of 81.3%. Figures 8.8 and 8.10 show the comparison of the performance of the baseline and FNDetector classifiers. Figures 8.9 and 8.10 show that both FNDetector_{RF} and FNDetector_{LSTM} models outperform the baseline classifiers in both accuracy and F-measure. We compute the improvements and present the results in the form of bar graphs (refer to Figures 8.10 and 8.11). We notice an improvement of 9.8% (accuracy) and 9.3% (F-measure) using FNDetector_{RF} model and an improvement of 17.7% (accuracy) and 40.2% (F-measure) using the FNDetector_{LSTM} model. These results indicate that feature extraction is an important step in fake news model creation. From the result, we can conclude that feature extraction is useful in fake news model building for both random forest and LSTM classifiers. We obtained an improvement of 9.8% (FNDetector_{RF}) and 17.7% (FNDetector_{LSTM}) in

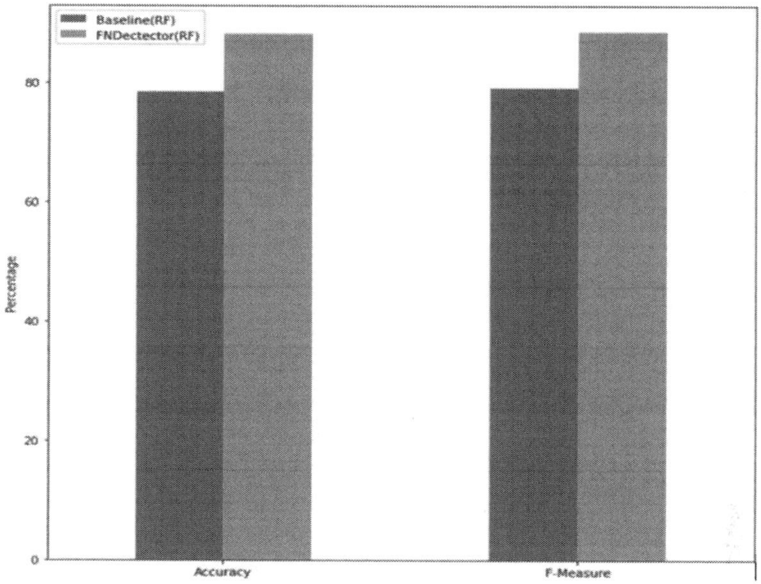

Figure 8.8 The comparison of the performances of baseline_{RF} classifier and FNDetector_{RF}.

terms of accuracy and an improvement of 9.3% (FNDetector_{RF}) and 40.2% (FNDetector_{LSTM}) in F-measure.

RQ 3: How important to extract features from both the title and contents of the article?

Motivation: Most of the researchers have focused on analyzing the headlines of the news articles for fake news model creation such as [66–68]. Some of the datasets that are used for building a fake news detection model such as the LIAR dataset [69]) do not contain both the title and contents of the news articles in the dataset. The dataset only contains the statement extracted from both the title and contents of the article. We believe that the content of news can provide important information for fake news detection. Thus, in this RQ, we work on analyzing the importance of both titles and contents for detecting fake news. The result of this RQ can help us understand which part of the news is more important in fake news detection.

Approach: To answer this RQ, we build three variants of the FNDetector model: FNDetector (title), FNDetector (content) and FNDetector (title+content). FNDetector (title), model is trained using only the title of the news articles. FNDetector (content), is trained using the content of the news articles. FNDetector (title+content) is trained using both title and content of

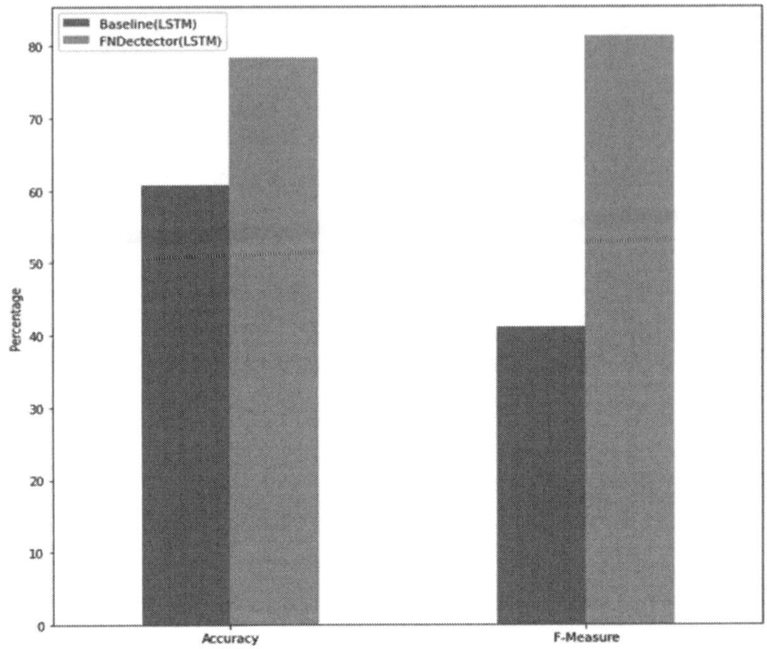

Figure 8.9 The comparison of the performances of $\text{baseline}_{\text{LSTM}}$ classifier and $\text{FNDetector}_{\text{LSTM}}$.

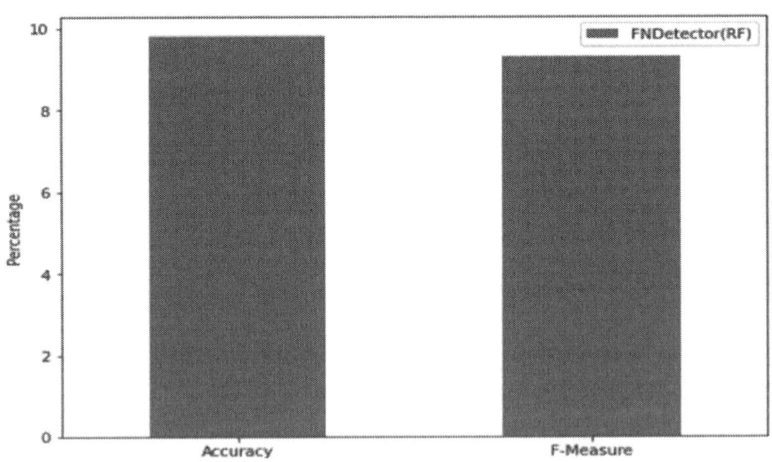

Figure 8.10 The improvement obtained in accuracy and F-measure using $\text{FNDetector}_{\text{RF}}$ compared to $\text{baseline}_{\text{RF}}$ classifier.

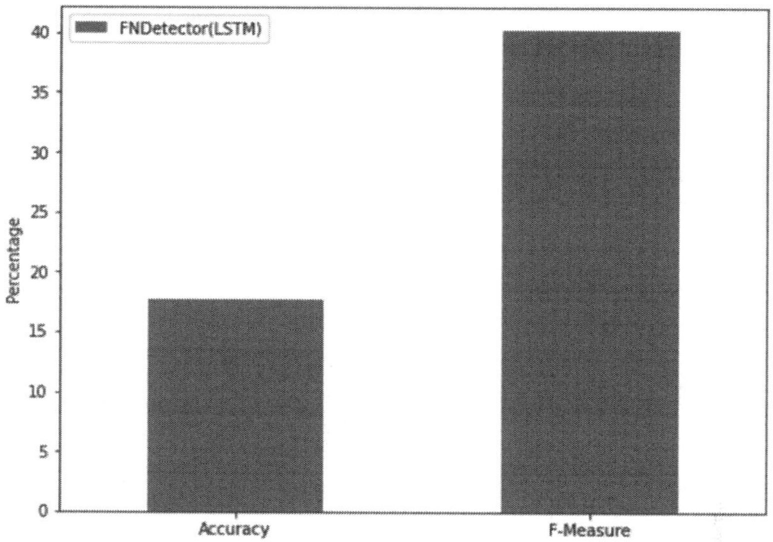

Figure 8.11 The improvement obtained in accuracy and F-measure using FNDetector_{LSTM} compared to baseline_{LSTM} classifier.

Table 8.13 The performance of the FNDetector model uses extracted features from the title of the news article

Method	*Accuracy*	*Precision*	*Recall*	*F-Measure*
FNDetector_{RF} (title)	70.6%	69.0%	76.9%	72.7%
FNDetector_{LSTM} (title)	74.5%	75.0%	77.8%	76.4%

the news articles which is the original FNDetector model is proposed in this work. We train all these variants using both RF and LSTM classifiers, hence, create a total of six variants of the FNDetector model.

Results: Tables 8.13, 8.14, and 8.15 show the results of this RQ using both RF and LSTM classifiers. Figures 8.12 and 8.13 show the comparison of the performance between different variants for FNDetector_{RF} and FNDetector_{LSTM} models, respectively. Table 8.13 shows that the accuracy and F-measure using FNDetector_{RF} (title) model is 70.6% and 72.7%, respectively. The FNDetector_{LSTM} (title) gives accuracy and F-measure of 74.5% and 76.4%, respectively. We observe from Table 8.16 that the performance of the FNDetector_{RF} model is improved when we use "content" of news article as compared to when we use "title" for training the model. We observe an improvement of 11.8% (accuracy) and 10.3% (F-measure) when using FNDetector_{RF} (content) as compared to the FNDetectorRF (title)

Table 8.14 The performance of the FNDetector model uses extracted features from the contents of the news article

Method	*Accuracy*	*Precision*	*Recall*	*F-Measure*
$\text{FNDetector}_{\text{RF}}$ (content)	82.4%	91.7%	75.9%	83.0%
$\text{FNDetector}_{\text{LSTM}}$ (content)	58.8%	87.5%	25.9%	40.0%

Table 8.15 The performance of the FNDetector model uses extracted features from both title and contents of the article

Method	*Accuracy*	*Precision*	*Recall*	*F-Measure*
$\text{FNDetector}_{\text{RF}}$ (title + content)	88.2%	85.2%	92.0%	88.5%
$\text{FNDetector}_{\text{LSTM}}$ (title + content)	78.4%	75.0%	88.9%	81.4%

Table 8.16 Improvement of FNDetectorRF (title+content)

	Imp. in Accuracy	*Imp. in F-Measure*
$\text{FNDetector}_{\text{RF}}$ (title)	17.6%	15.8%
$\text{FNDetector}_{\text{RF}}$ (content)	5.8%	5.5%

Table 8.17 Improvement of $\text{FNDetector}_{\text{LSTM}}$ (title+content)

	Imp. in Accuracy	*Imp. in F-Measure*
$\text{FNDetector}_{\text{LSTM}}$ (title)	3.9%	5.0%
$\text{FNDetector}_{\text{LSTM}}$ (content)	19.6%	41.4%

model. For the LSTM classifier, we observe a reduction in the performance in both accuracy and F-measure when the "content" of news articles is used to train the classifier (refer to Table 8.17).

We notice an enhancement in the model's performance when we use both the title and content of the news article to train the model (refer to Table 8.15, Figures 8.12, and 8.13). This is also the original FNDetector model proposed in the work. Figures 8.12 and 8.13 show a comparison of the performances of the three variants for $\text{FNDetector}_{\text{RF}}$ and $\text{FNDetector}_{\text{LSTM}}$ models, respectively. We notice an improvement in the performances of both $\text{FNDetector}_{\text{RF}}$ and $\text{FNDetector}_{\text{LSTM}}$ models. Using $\text{FNDetector}_{\text{RF}}$ (title + content), we notice an improvement of 17.6% (in comparison to $\text{FNDetector}_{\text{RF}}$ (title)) and 5.8% (in comparison to $\text{FNDetector}_{\text{RF}}$ (content)) in accuracy. Similarly, using $\text{FNDetector}_{\text{LSTM}}$ (title+content), we notice an improvement

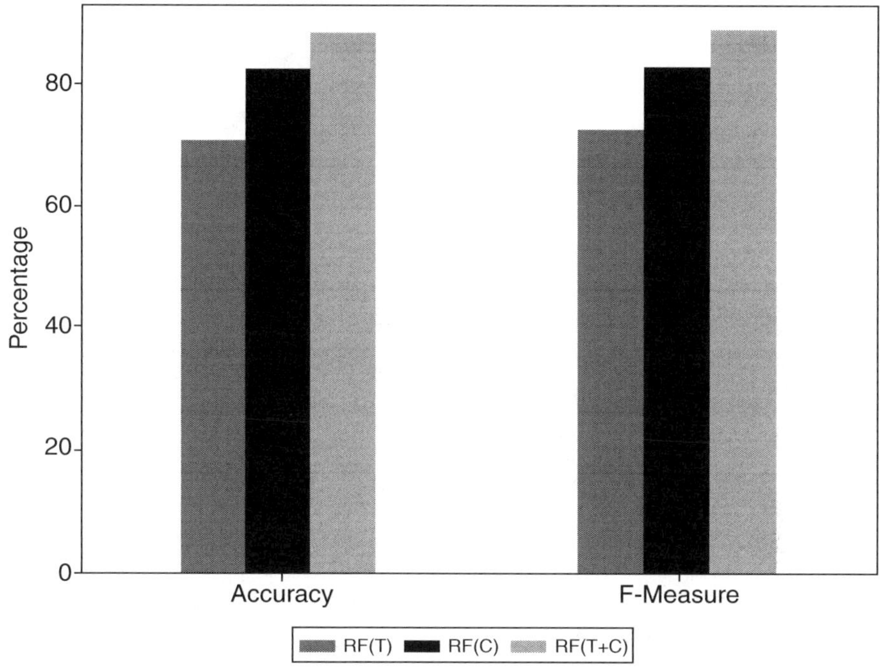

Figure 8.12 Comparison between FNDetector models in random forest classifier with title, content and title+content. T: Title, C: Content.

of 3.93% (in comparison to $\text{FNDetector}_{\text{LSTM}}$ (title)) and 19.63% (in comparison to $\text{FNDetector}_{\text{LSTM}}$ (content)) in accuracy. The improvements are also shown in Tables 8.16 and 8.17. Results indicate that both the title and content of news articles are important in fake news detection. For both $\text{FNDetector}_{\text{RF}}$ and $\text{FNDetector}_{\text{LSTM}}$, we observed a significant improvement in the prediction accuracy when both title and content are used to train the model.

RQ 4: Which features are more important to detect fake news in the FNDetector model?

Motivation: We use 32 features to train the FNDetector model. These features belong to different domains and have different types, i.e., linguistic-based features (grammatical and lexical features), sentiment-based features, named entity recognition features, and textual features. The FNDetector model outdid the baseline line classifier significantly in both accuracy and F-measure. An analysis of the performance of the proposed FNDetector model with incremental feature class addition can provide insights into the importance of different feature classes.

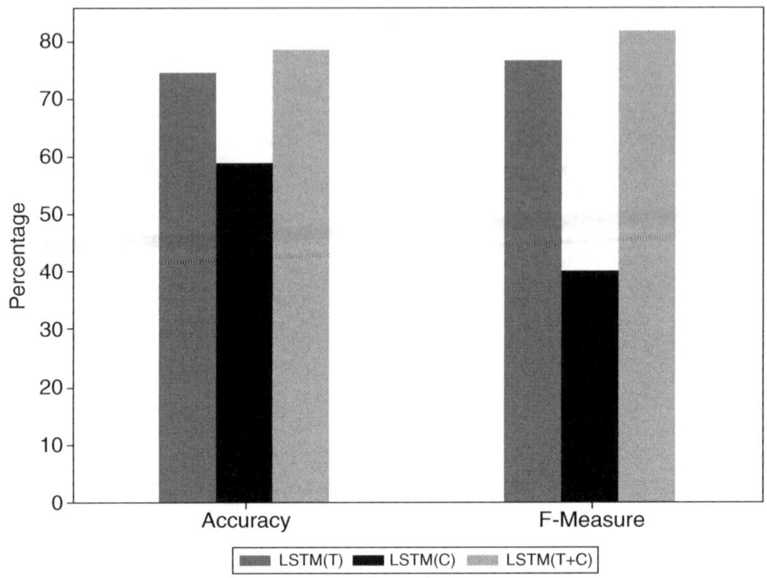

Figure 8.13 Comparison between FNDetector models in LSTM classifier with title, content and title+content. T: Title, C: Content.

Table 8.18 Notation for each feature set

Feature name	*Feature symbol*
Linguistic-based feature	FL
Sentiment-based feature	FS
Named entity recognition	FN
Textual features	FT

Approach: To answer this RQ, we perform two types of experiments. First, we created four variants of FNDetector models (for both RF and LSTM classifiers) and added features incrementally to observe the effect of each feature class on the performance of the FNDetector model. We create the following variants:

- $\text{FNDetector}_{A}^{F_S}$: FNDetector model trained using sentiment-based features.
- $\text{FNDetector}_{A}^{F_S+F_L}$: FNDetector model trained using sentiment-based and linguistics-based features
- $\text{FNDetector}_{A}^{F_S+F_L+F_N}$: FNDetector model trained using sentiment-based, linguistics-based features, and NER-based features.

$FNDetector_{A}^{F_S+F_L+F_N+F_T}$: FNDetector model trained using all the 32 features extracted.

In the above, A refers to the classifier. A can take value RF or LSTM. Hence, we compute the performances of all the above variants with both RF and LSTM classifiers. Second, we compute the feature importance of all the numerical features using an RF classifier. Table 8.18 shows the notation for each feature set.

Results: Figures 8.14, 8.15, 8.16, and Table 8.19 show the result of the FNDetector model using different feature sets. We added features from different features incrementally. Table 8.19 shows a consistent improvement in the performance (for both accuracy and F-measure) of the FNDetector model with the addition of each feature set. In the next experiment, we calculate each numerical feature's relevance using the RF classifier.

The RF classifier has the advantage of getting the feature importance from the training model. The feature importance using the RF classifier has been presented in Figure 8.16. From the result, it is seen that the quantity of nouns (noun title and noun content) in both the title and contents of the news article has the most impact on the training model (when using the RF classifier). Using the results of this RQ, we can observe that all the feature sets used

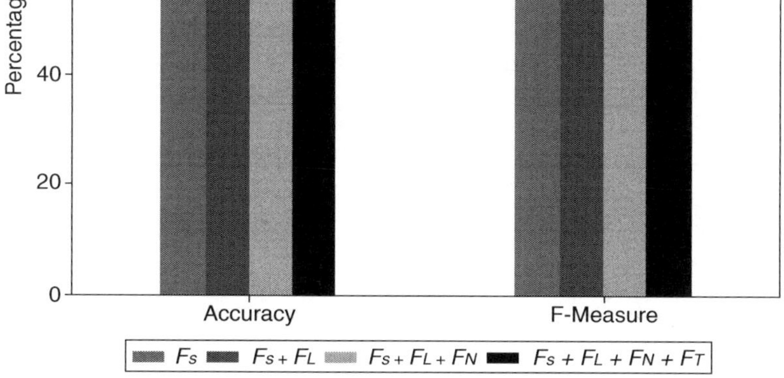

Figure 8.14 Comparison of performance based on different extracted features in Random Forest Classifier (For notations refer to Table 8.18).

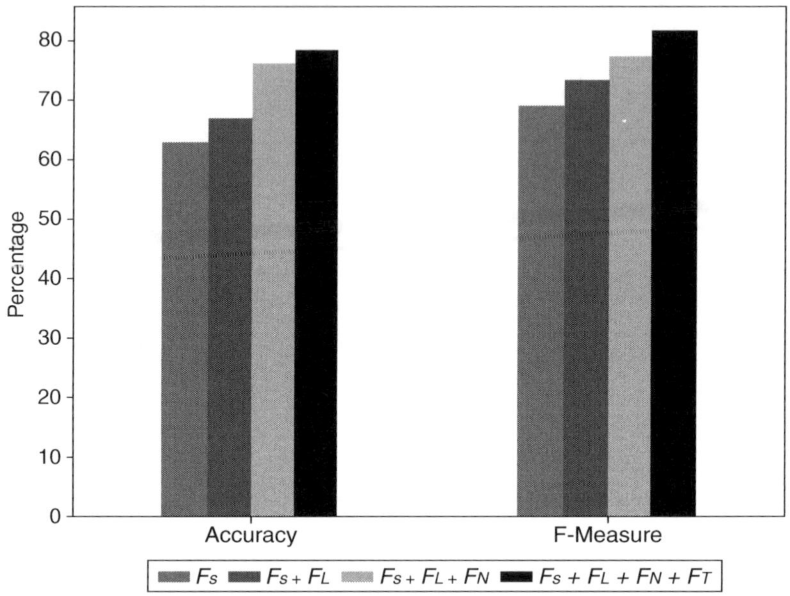

Figure 8.15 Comparison of performance based on different extracted features in LSTM Classifier (For notations refer to Table 8.18).

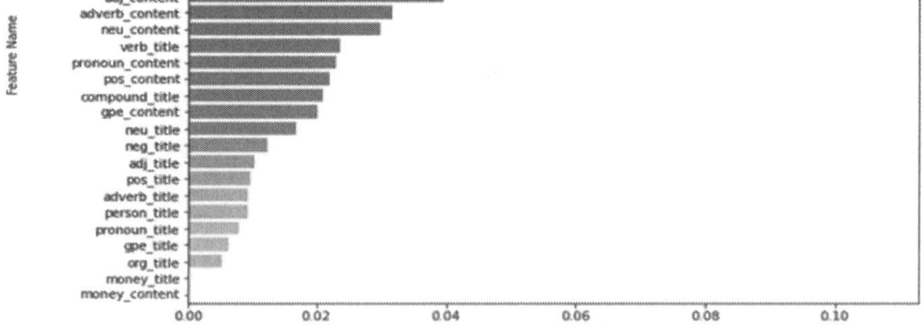

Figure 8.16 Feature importance of random forest classifier.

Table 8.19 The performance of FNDetector model with different feature class set

	RF Classifier (%)	*LSTM Classifier (%)*	*Feature Sets*
Accuracy	64.7	62.7	$FNDetector_{A}^{F_S}$
Precision	59.3	61.2	
Recall	69.6	77.8	
F-Measure	64.0	68.9	
Accuracy	72.5	66.7	$FNDetector_{A}^{F_S+F_L}$
Precision	78.9	63.9	
Recall	60.0	85.1	
F-Measure	68.2	73.0	
Accuracy	76.5	70.6	$FNDetector_{A}^{F_S+F_L+F_N}$
Precision	76.0	65.8	
Recall	76.0	92.6	
F-Measure	76.0	76.9	
Accuracy	88.2	78.4	$FNDetector_{A}^{F_S+F_L+F_N+F_T}$
Precision	85.2	75.0	
Recall	92.0	88.9	
F-Measure	88.5	81.4	

to train the model are important and help in improving the accuracy of the FNDetector model. noun title, noun content, totalWord title, and totalWord content features were found to have the highest impact on the performance of the FNDetector model (when trained using the RF classifier).

RQ 5: What is the effect on the performance of FNDetector model by changing parameters of RF and LSTM classifiers?

Motivation: The parameters of an algorithm play an important role as they can impact the performance of an algorithm. It's critical to adjust an algorithm's settings., in order to build a high-accuracy model. In this RQ, we will present what is the effect on the performance of the FNDetector model by changing the parameters of RF and LSTM classifiers. We also compute the time it requires to train the models as well as the time required to make the predictions. We believe that time can be an important factor in choosing the appropriate classifier in resource-constrained applications like mobile devices.

Approach: To answer the RQ, we perform several experiments with the FNDetector model using RF and LSTM variants. We create a total of seven variants of the FNDetector model, three using $FNDetcetor_{RF}$ and four using $FNDetector_{LSTM}$. For $FNDetector_{RF}$, we tested the performance of the model for three different values of the estimator, i.e., 50, 100, and 150. For $FNDetector_{LSTM,}$ we tested the performance of the model by changing

Table 8.20 The parameters that will be changed in experiments

Method	*Parameter*	*Value*
	n estimators	50
$FNDetector_{RF}$		100
		150
	epoch	5
$FNDetector_{LSTM}$		10
	batch Size	128
		256

epochs and batch size. We use 2 different epochs, i.e., 5 and 10 epochs, and 2 different batch sizes, i.e., 128 and 256 respectively. Table 8.20 shows the parameters and values that we use to perform the experiments.

The time required to train the model and generate a prediction was then calculated. We compute the time it takes to build the model using the training dataset for both $FNDetector_{RF}$ and $FNDetector_{LSTM}$ classifiers for different parameters (number of estimators (for RF classifier) and epoch & batch size (for LSTM classifier)) as discussed above.

Results: Figures 8.17, 8.18, and 8.19 show the result of $FNDetector_{RF}$ classifier with 50, 100, and 150 n estimators. From the result, we can observe count of FPs and FNs decreases when the n_estimators increase. Also, the AUC value is increased when the n_estimators are increased. These results show that increasing the number of estimators can aid in enhancing the performance of $FNDetector_{RF}$ (please refer to Table 8.21). For $FNDetector_{LSTM}$ classifier, we demonstrated the model accuracy and loss of the model by plotting it in Figures 8.20 and 8.21. The results show that the $FNDetector_{LSTM}$ classifier attained the best result with epoch 10 and batch size 128. We can observe that the loss is decreasing and the accuracy of the model is increasing consistently as the number of epochs increases. This makes the $FNDetector_{LSTM}$ model with 10 epochs and 128 batch sizes have the highest accuracy in detecting fake news compared to other parameters.

Table 8.21 shows that for $FNDetector_{RF}$ classifier the training and prediction time increases as we increase the number of estimators. As we are increasing the number of estimators the RF algorithm needs to create more trees and hence, the time increases. For the $FNDetector_{LSTM}$ classifier, we observe a similar pattern, i.e., training time increase as we intensify the number of epochs. As we increase, batch size the training time decrease for the $FNDetector_{LSTM}$ classifier. This happened because with increases in batch size the algorithm can take big chunks of the training dataset and can learn faster. Overall, we observe that the $FNDetector_{RF}$ classifier took much less time in both training and testing as compared to the $FNDetector_{LSTM}$ classifier. Experimental results show the parameter tuning is very important as

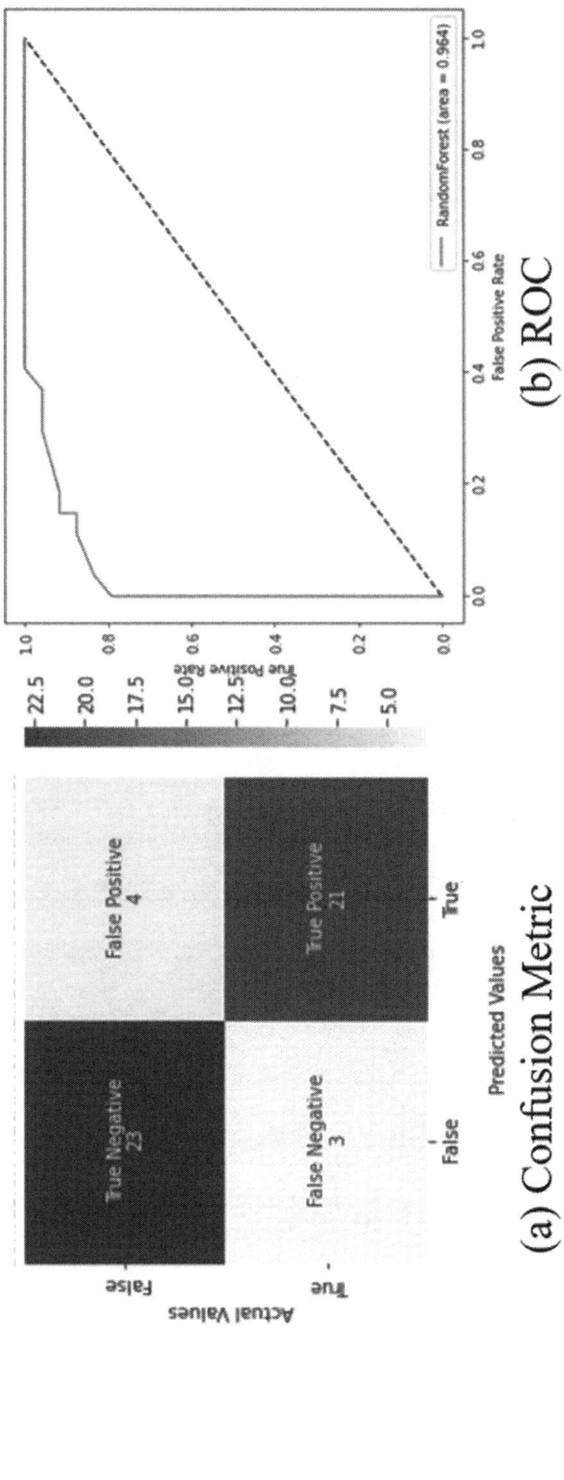

Figure 8.17 Confusion metric and ROC plot for FNDetectorRF classifier with n estimator = 50.

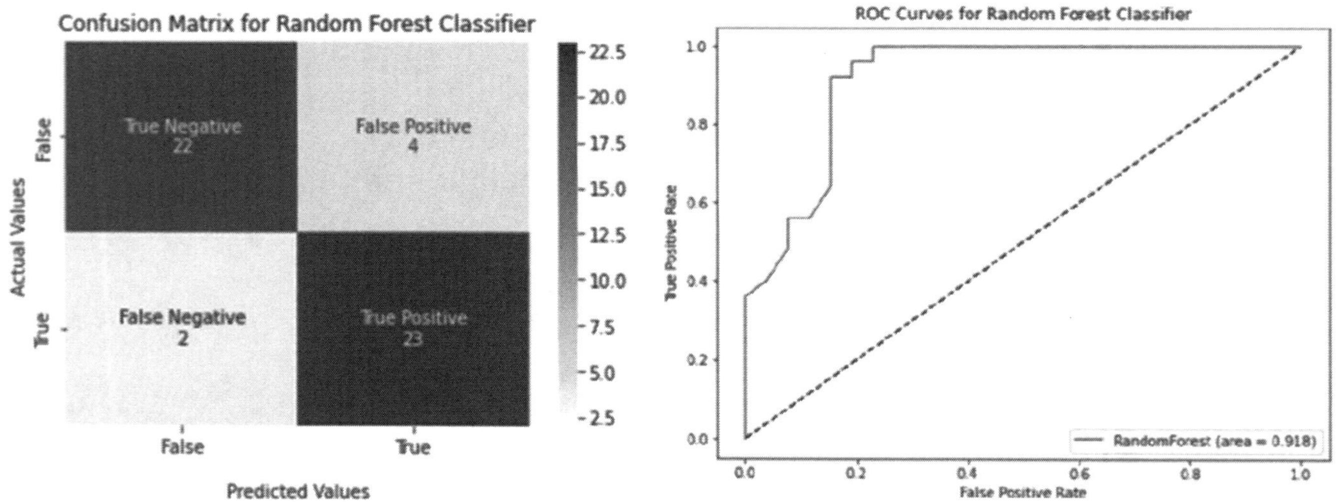

(a) Confusion Metric

(b) ROC

Figure 8.18 Confusion metric and ROC plot for FNDetectorRF classifier with n estimator = 100.

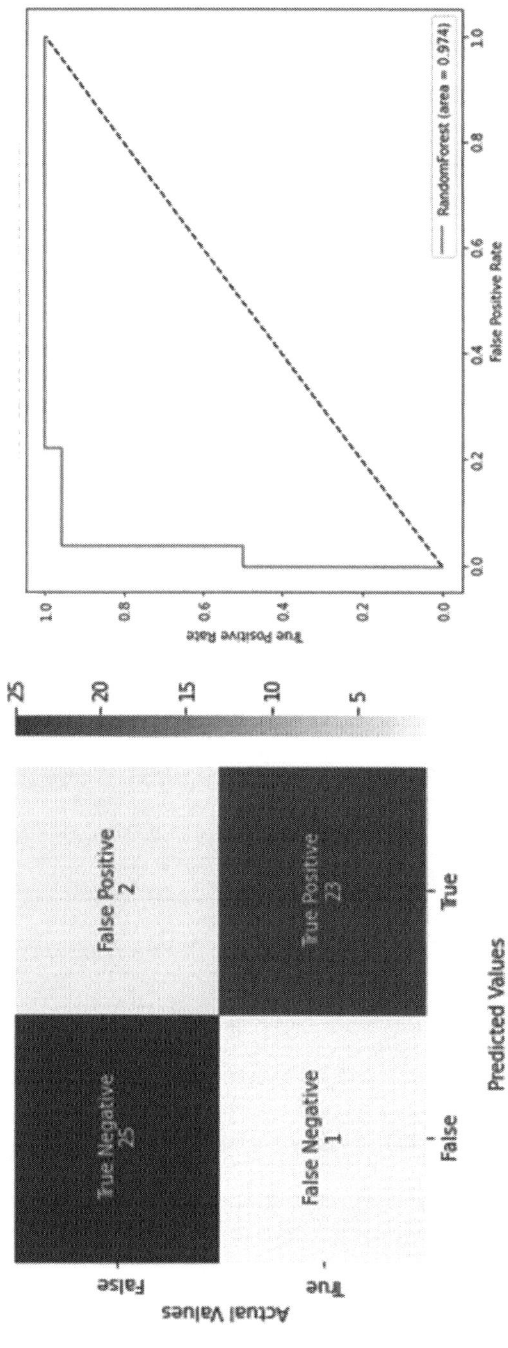

(a) Confusion Metric

(b) ROC

Figure 8.19 Confusion metric and ROC plot for FNDetectorRF classifier with n estimator = 150.

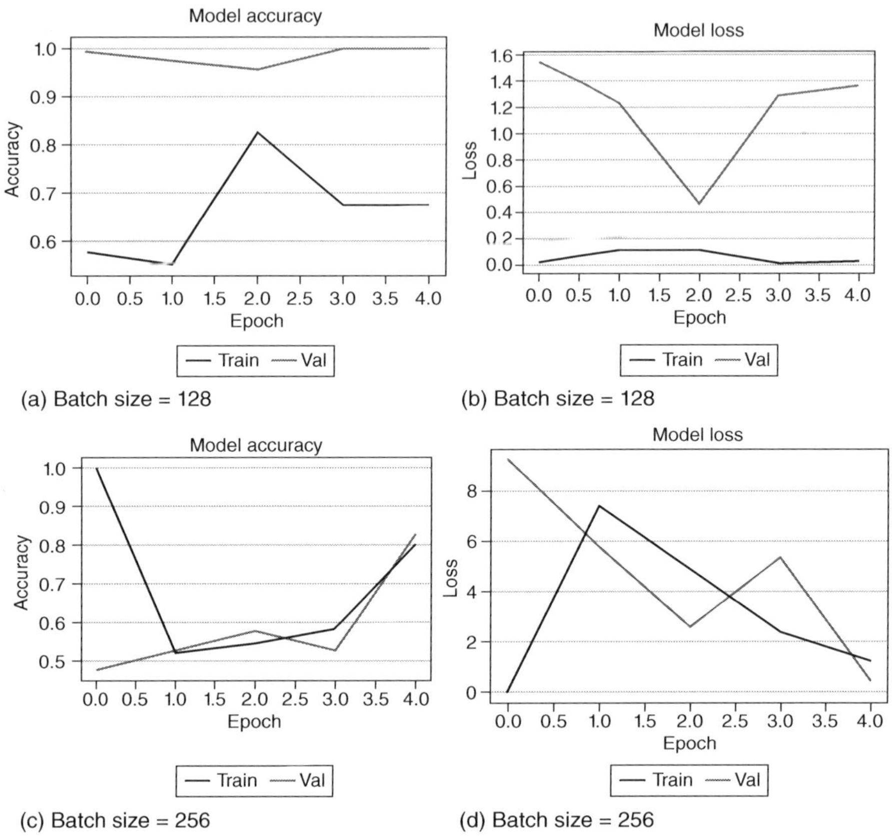

Figure 8.20 Accuracy and loss plot for $\text{FNDetector}_{\text{LSTM}}$ classifier with epoch = 5.

it can have a big influence FNDetector model's effectiveness. In our experiment, the RF classifier achieved its best performance when the n estimator was set to 150 and the LSTM classifier achieve its best performance when epochs value is 10 and the batch size value is 128. $\text{FNDetector}_{\text{RF}}$ took much less training and testing time as compared to the $\text{FNDetector}_{\text{LSTM}}$ classifier.

8.9 CONCLUDING REMARKS AND FUTURE SCOPE

The amount of news items shared on social networking websites has skyrocketed. Because so much false information circulates on social media, people are less likely to believe news articles published on these channels. It's challenging to confirm the veracity of news articles on social media. The identification of fake news has become more significant as a way to get around this problem and as a motivator for related research. We provide

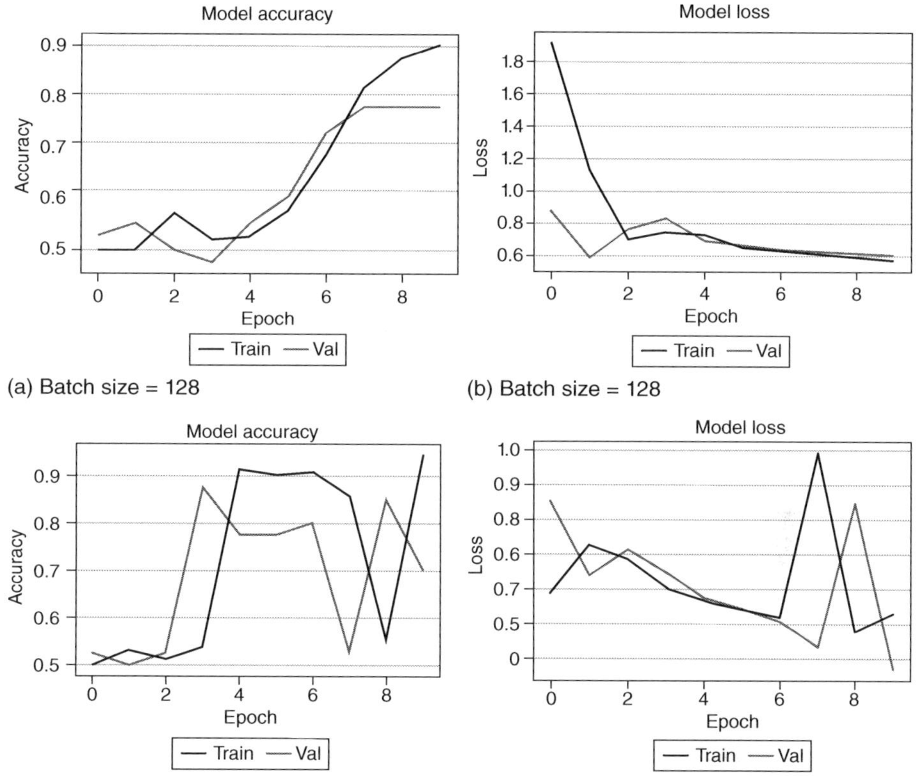

(a) Batch size = 128 (b) Batch size = 128

(c) Batch size = 256 (d) Batch size = 256

Figure 8.21 Accuracy and loss plot for FNDetector$_{LSTM}$ classifier with epoch = 10.

Table 8.21 Total training and prediction time of FNDetector$_{RF}$ and FNDetector$_{LSTM}$ with different parameters

Method	*Parameter*	*Value*	*Training Time (secs)*	*Prediction Time(secs)*
FNDetector$_{RF}$	n estimators	50	0.30	0.03
		100	0.33	0.04
		150	0.40	0.04
FNDetector$_{LSTM}$	epoch/batch size	5/128	25.0	2.39
		5/256	16.0	2.24
		10/128	53.0	2.24
		10/256	31.0	2.53

FNDetector, a fake news detection model, in this paper. FNDetector uses a combination of various feature sets. We train the FNDetector using two classifiers, i.e., RF and LSTM. The FNDetector model outperformed the baseline classifier significantly. $FNDetector_{RF}$ and $FNDetector_{LSTM}$ give an accuracy of 88.2% and 78.4%, respectively. Our investigation related to feature importance shows that all the feature sets used in this paper are important and help in increasing the accuracy. Also, 'noun count' and 'total word count' in both the title and content of news articles are found to be the top four important features. We intend to use ensemble learning to expand the model in the future. The ensemble learning model combines different classifiers to improve training models. The ensemble learning model working has been shown in a few studies such as [69] and [45] to improve the outcomes of the models. Additionally, we plan to use more advanced methods such as BERT to enhance the effectiveness of fake news detection.

REFERENCES

[1] Cambridge dictionary. (n.d). Cambridge dictionary. https://dictionary.cambridge.org/ dictionary/english/fake-news

[2] BBC. (2022). A brief history of fake news. www.bbc.co.uk/bitesize/articles/zwcgn9q. [Accessed Date: 22 June 2022].

[3] Vosoughi, S., Roy, D., & Aral, S. (2018). The spread of true and false news online. *Science*, 359, 1146–1151. doi:10.1126/science.aap955

[4] Allcott, H., & Gentzkow, M. (2017). Social media and fake news in the 2016 election. *Journal of Economic Perspectives*, 31, 212. doi:10.1257/jep.31.2.211

[5] WHO (2022). Covid-19 mythbusters. www.who.int/emergencies/diseases/novel-coronavirus-2019/advice-for-public/myth-busters#pepper [Accessed Date: 22 June 2022].

[6] BBC (2020). Coronavirus: Fake and misleading stories that went viral this week. www.bbc.co.uk/news/52124740. [Accessed Date: 28 July 2022].

[7] de Beer, D., & Matthee, M. (2020). Approaches to identify fake news: A systematic literature review. *Integrated Science in Digital Age* 2020, (p. 13–22). doi:10.1007/978-3-030-49264-9_2

[8] Wang, W. Y. (2017). "liar, liar pants on fire": A new benchmark dataset for fake news detection. *arXiv preprint arXiv:1705.00648*.

[9] Choudhary, A., & Arora, A. (2021). Linguistic feature-based learning model for fake news detection and classification. *Expert Systems with Applications*, 169, 1–15. doi:10.1016/j.eswa.2020. 114171

[10] Lischka, J. A., & Garz, M. (2021). Clickbait news and algorithmic curation: A game theory framework of the relation between journalism, users, and platforms. New Media and Society, (p. 3). https://journals.sagepub.com/doi/full/10.1177/14614448211027174#bibr56-14614448211027174. doi:10.1177/14614448211027174

[11] Hu, X. E. (2017). Clickbait and conscientiousness. https://stanforddaily.com/2017/03/ 20/clickbait-and-conscientiousness/ [Accessed Date: 23 June 2022].

[12] Barr´on-Cedeno, A., Da San Martino, G., Jaradat, I., & Nakov, P. (2019). Proppy: A system to unmask propaganda in online news. In *Proceedings of the AAAI Conference on Artificial Intelligence* (pp. 9847–9848). volume 33.

[13] Lewis, H. (2016). How the Brexit campaign lied to us–and got away with it. New Statesman, 30. www.newstatesman.com/politics/2016/06/ how-brexit-campaign-lied-us-and-got-away-it [Accessed Date: 27 June 2022].

[14] The Independent (2018). The misinformation that was told about brexit during and after the referendum. www.independent.co.uk/news/uk/politics/ final-say-brexit-referendum-lies-boris-johnson-leave-campaign-remain-a8466751. [Accessed Date: 23 June 2022].

[15] Shu, K., Sliva, A., Wang, S., Tang, J., & Liu, H. (2017). Fake news detection on social media. *ACM SIGKDD Explorations Newsletter* , 19, 22–36. doi:10.1145/3137597.3137600.

[16] BuzzFeedNews Dataset. https://github.com/BuzzFeedNews/2016-10-facebook-fact-check/tree/master/data

[17] BuzzFeedNews Dataset. www.cs.ucsb.edu/~william/data/liar_dataset.zip

[18] BS Detector Dataset. www.kaggle.com/mrisdal/fake-news

[19] CREDBANK Dataset. http://compsocial.github.io/CREDBANK-data/

[20] Bhutani, B., Rastogi, N., Sehgal, P., & Purwar, A. (2019). Fake news detection using sentiment analysis. In 2019 Twelfth International Conference on Contemporary Computing (IC3) (pp. 1–5). doi:10.1109/IC3.2019.8844880

[21] George McIntire Dataset. https://github.com/GeorgeMcIntire/fake_real_news_dataset

[22] Kaggle Dataset. www.kaggle.com/mrisdal/fake-news

[23] PolitiFact Dataset. www.kaggle.com/datasets/rmisra/politifact-fact-check-dataset

[24] Khan, S., Hakak, S., Deepa, N., Prabadevi, B., Dev, K., & Trelova, S. (2022). Detecting covid-19-related fake news using feature extraction. *Frontiers in Public Health*, 9. doi:10.3389/fpubh. 2021.788074

[25] Iwendi, C., Mohan, S., khan, S., Ibeke, E., Ahmadian, A., & Ciano, T. (2022). Covid-19 fake news sentiment analysis. *Computers and Electrical Engineering*, 101, 1–22. doi:10.1016/j. compeleceng.2022.107967

[26] Ajao, O., Bhowmik, D., & Zargari, S. (2019). Sentiment aware fake news detection on online social networks. *In ICASSP 2019–2019 IEEE International Conference on Acoustics, Speech and Signal Processing (ICASSP)* (pp. 2507–2511). doi:10.1109/ICASSP.2019.8683170

[27] Reis, J. C. S., Correia, A., Murai, F., Veloso, A., & Benevenuto, F. (2019). Supervised learning for fake news detection. *IEEE Intelligent Systems*, 34, 76–81. doi:10.1109/MIS.2019.2899143

[28] Khanam, Z., Alwasel, B. N., Sirafi, H., & Rashid, M. (2021). Fake news detection using machine learning approaches. *IOP Conference Series: Materials Science and Engineering*, 1099, 1–13. doi:10.1088/1757-899x/1099/1/012040

[29] Mishra, S., Shukla, P., & Agarwal, R. (2022). Analyzing machine learning enabled fake news de- tection techniques for diversified datasets. *Wireless Communications and Mobile Computing*, 2022, 1–18. doi:10.1155/2022/1575365

[30] Khan, J. Y., Khondaker, M. T., Afroz, S., Uddin, G., & Iqbal, A. (2021). A benchmark study of machine learning models for online fake news detection. *Machine Learning with Applications*, 4, 1–12. doi:10.1016/j.mlwa.2021.100032

[31] Devlin, J., Chang, M.-W., Lee, K., & Toutanova, K. (2019). Bert: Pre-training of deep bidirectional transformers for language understanding. In *Proceedings of the 2019 conference of the North American chapter of the association for computational linguistics: human language technologies, volume 1 (long and short papers)*, pp. 4171–4186.

[32] Paul, S., Joy, J. I., Sarker, S., Shakib, A. A. H., Ahmed, S., & Das, A. K. (2019). Fake news detection in social media using blockchain. In 2019 7th International Conference on Smart Computing and Communications (ICSCC) (pp. 1–5). doi:10.1109/ICSCC.2019.8843597

[33] Flair library. https://github.com/zalandoresearch/flair

[34] Akbik, A., Blythe, D., & Vollgraf, R. (2018). Contextual string embeddings for sequence labeling. In COLING 2018, 27th International Conference on Computational Linguistics (pp. 1638–1649).

[35] Kula, S., Chora´s, M., Kozik, R., Ksieniewicz, P., & Wo´zniak, M. (2020). Sentiment analysis for fake news detection by means of neural networks. *Lecture Notes in Computer Science*, 12140, 653–666. doi:10.1007/978-3-030-50423-6_49

[36] Horne, B. D., & Adali, S. (2017). This just in: Fake news packs a lot in title, uses simpler, repetitive content in text body, more similar to satire than real news. In Eleventh international AAAI conference on web and social media.

[37] Balshetwar, S. V., Rs, A., and D. J. R. (2023) Fake news detection in social media based on sentiment analysis using classifier techniques. *Multimedia Tools and Applications*, 82, no. 23, 35781–35811.

[38] Uppada, S. K., Patel, P., and Sivaselvan, B. (2023). An image and text-based multimodal model for detecting fake news in OSN's. *Journal of Intelligent Information Systems*, 61, 367–393. doi:10.1007/s10844-022-00764-y

[39] Balakrishnan, V., & Lloyd-Yemoh, E. (2014). Stemming and lemmatization: A comparison of retrieval performances. *Lecture Notes on Software Engineering* 2, no. 3, 262–267.

[40] Manning, C. D., Raghavan, P., & Hinrich, S. (2008). *Introduction to information retrieval.* Cambridge University Press.

[41] Princeton University (2010). About wordnet. https://wordnet.princeton.edu/ [Accessed Date: 29 July 2022].

[42] Mahyoob, M., Al-Garaady, J., & Alrahaili, M. (2020). Linguistic-based detection of fake news in social media. *International Journal of English Linguistics*, 11, 99–109.

[43] Conroy, N. K., Rubin, V. L., & Chen, Y. (2015). Automatic deception detection: Methods for finding fake news. *Proceedings of the Association for Information Science and Technology*, 52, 1–4. doi:10.1002/pra2.2015.145052010082

[44] Chiche, A., & Yitagesu, B. (2022). Part of speech tagging: A systematic review of deep learning and machine learning approaches. *Journal of Big Data*, 9. doi:10.1186/s40537-022-00561-y

[45] Alonso, M. A., Vilares, D., G´omez-Rodr´ıguez, C., & Vilares, J. (2021). Sentiment analysis for fake news detection. *Electronics*, 10, 1348. doi:10.3390/electronics10111348

[46] Martel, C., Pennycook, G., & Rand, D. G. (2020). Reliance on emotion promotes belief in fake news. *Cognitive Research: Principles and Implications*, 5, 1–20.

[47] (NLTK) package. www.nltk.org/

[48] Shelar, H., Kaur, G., Heda, N., & Agrawal, P. (2020). Named entity recognition approaches and their comparison for customer model. *Science and Technology Libraries*, 39, 324–337. doi:10.1080/0194262x.2020.1759479

[49] Qaiser, S., & Ali, R. (2018). Text mining: Use of tf-idf to examine the relevance of words to documents. *International Journal of Computer Applications*, 181, 25–29. doi:10.5120/ ijca2018917395

[48] scikit-learn package. https://scikit-learn.org/stable/modules/feature_extraction.html

[51] TfidfVectorizer.html Tokenizer package. www.tensorflow.org/api_docs/python/tf/keras/preprocessing/text/Tokenizer

[52] Breiman, L. (2001). Random forests. *Machine Learning*, 45, 5–32. doi:10.1023/a:1010933404324

[53] Lee, T.-H., Ullah, A., & Wang, R. (2020). Bootstrap aggregating and random forest. In Macroeconomic forecasting in the era of big data (pp. 389–429). Springer.

[54] Song, J., Gao, Y., Yin, P., Li, Y., Li, Y., Zhang, J., Su, Q., Fu, X., & Pi, H. (2021). The random forest model has the best accuracy among the four pressure ulcer prediction models using machine learning algorithms. *Risk Management and Healthcare Policy*, 14, 1175–1187. doi:10.2147/rmhp.s297838

[55] Polamuri, S. R., Srinivas, K., & Mohan, A. K. (2019). Stock market prices prediction using random forest and extra tree regression. *International Journal of Recent Technology and Engineering*, 8, 1224–1228. doi:10.35940/ijrte.c4314.098319

[56] Hochreiter, S., & Schmidhuber, J. (1997). Long short-term memory. *Neural Computation*, 9, 1735–1780. doi:10.1162/neco.1997.9.8.1735

[57] Song, X., Liu, Y., Xue, L., Wang, J., Zhang, J., Wang, J., Jiang, L., & Cheng, Z. (2020). Time-series well performance prediction based on long short-term memory (lstm) neural network model. *Journal of Petroleum Science and Engineering*, 186. doi:10.1016/j.petrol.2019.106682

[58] Lindemann, B., Mu¨ller, T., Vietz, H., Jazdi, N., & Weyrich, M. (2021). A survey on long short- term memory networks for time series prediction. *Procedia CIRP*, 99, 650–655. doi:10.1016/ j.procir.2021.03.088

[59] Python package. www.python.org/

[60] Caruana, R., & Niculescu-Mizil, A. (2004). Data mining in metric space: an empirical analysis of supervised learning performance criteria. In Proceedings of the tenth ACM SIGKDD international conference on Knowledge discovery and data mining (pp. 69–78).

[61] Hossin, M. & Sulaiman, Md N. (2015). "A review on evaluation metrics for data classification evaluations." *International Journal of Data Mining & Knowledge Management Process*, 5, no. 2, 1

[62] Joshi, M. (2002). On evaluating performance of classifiers for rare classes. In 2002 IEEE International Conference on Data Mining, 2002. Proceedings. (pp. 641–644). doi:10.1109/ICDM. 2002.1184018

[63] Hajian-Tilaki, K. (2013). Receiver operating characteristic (roc) curve analysis for medical diagnostic test evaluation. *Caspian Journal of Internal Medicine*, 4, 627.

[64] Park, S. H., Goo, J. M., & Jo, C.-H. (2004). Receiver operating characteristic (roc) curve: practical review for radiologists. *Korean Journal of Radiology*, 5, 11–18.

[65] Mandrekar, J. N. (2010). Receiver operating characteristic curve in diagnostic test assessment. *Journal of Thoracic Oncology*, 5, 1315–1316. doi:10.1097/jto.0b013e3181ec173d

[66] Sisodia, D. S. (2019). Ensemble learning approach for clickbait detection using article headline features. *Informing Science*, 22, 31.

[67] Liu, H., He, D., & Chan, S. (2021). Fraudulent news headline detection with attention mechanism. *Computational Intelligence and Neuroscience*, 2021, 1–7. doi:10.1155/2021/6679661

[68] Eembi, N. C., Ishak, I., Sidi, F., Affendey, L. S. et al. (2017). Towards a headline-based deception detection approach for data veracity in online digital news. *Journal of Telecommunication, Electronic and Computer Engineering (JTEC)*, 9, 33–37.

[69] Xia, X., Lo, D., Shihab, E., Wang, X., & Yang, X. (2015). Elblocker: Predicting blocking bugs with ensemble imbalance learning. *Information and Software Technology*, 61, 93–106.

Chapter 9

Survey of visual deepfake detection methods

Nikhil Rathoure, R. K. Pateriya, Deepak Singh Tomar, Ankit Vidyarthi, Nitesh Bharot, and Priyanka Verma

9.1 INTRODUCTION

With the recent advancements in deep learning techniques, the world of a generation of hyper-realistic media generation and manipulation has seen unprecedented growth, and the impact it brought along with itself is tremendous and far-reaching.

Over the last few years, the use of AI-generated content has substantially increased on the internet. Especially during and after the COVID-19 pandemic period, the dependence of the public on digital media has increased multifold, with people spending more and more time in the virtual world. As the dependence on digital media has increased, misinformation through fake media has also increased significantly. One of the biggest sources of misinformation on the internet is "Deepfake", which derives its name from "deep learning" and "fake", referring to media synthesized and/or manipulated by deep learning methods. Deepfakes are hyper-realistic audio, video, or image that easily deceive humans. Deepfake is one of the most bothersome applications of AI in recent years in terms of its potential use for cybercrimes and has been called the most serious AI crime threat [1].

Damages resulting from the spread of misinformation through deepfakes encompass damage to financial resources, psychological/mental health harm, and damage to the reputation and trust of companies/celebrities/politicians, which is often irrecoverable, causing trust issues and chaos and unrest in society. According to a report by Sumsub [2], there has been an increase of 245% year on year in deepfakes worldwide from Q1 2023 till Q1 2024, with multifold growth in countries where elections are being held or scheduled in 2024. It reported YOY increase of deepfake incidents in India (+280%), USA (+303%), Mexico (+500%), Indonesia (+1550%), and South Korea (+1625%). The magnitude and spread of deepfakes through the increasing pervasiveness of digital media in our personal and professional lives have made deepfakes a de facto tool for misinformation in today's time. As a result, the field of deepfake detection has gathered the attention of researchers across industries, academics, and the government.

DOI: 10.1201/9781003569619-9

Thus, it has become important to analyze deepfake and related advances in deepfake generation methods and development in detection methods. In this review, we will focus on visual deepfake detection methods.

This article is organized as follows: Section 9.1 provides the introduction of deepfakes. Section 9.2 discusses the various datasets available for deepfake detection methods for training and testing. Section 9.3 gives in-depth detail of various existing deepfake detection methods. Metrics used for performance assessment are discussed in Section 9.4. Section 9.5 presents the various challenges related to deepfakes detection that still exist. Section 9.6 summarizes and concludes the article.

9.1.1 Types of Deepfakes

Deepfakes can be divided into two broad categories: Visual Deepfakes and audio Deepfakes.

Visual Deepfakes refer to images or videos that are created by synthesizing fake images/videos or manipulating real images or videos. These visual Deepfakes involve images or videos in which the original face in the source has been replaced with someone else face (Face Swap). Source facial expressions have been changed (Face Reenactment) or certain characteristics or attributes of the face have been changed (Facial attribute manipulation) or complete fake has been generated (Entire Face Synthesis) or movement of the subject in the video has been changed in ways that actually never happened (puppet).

Audio Deepfakes refer to audio or audiovisual media instances in which the complete fake voice has been synthesized (voice synthesis using techniques such as text-to-speech), or lip-syncing has been done by changing the voice of the subject in a video with another voice recording (voice swapping).

9.1.2 Categories of visual Deepfake generation methods

Visual Deepfake algorithms can be divided into four major categories (Figure 9.2): (1) face swap, (2) face reenactment, (3) entire face synthesis, and (4) facial attributes manipulation. These major categories of Visual deepfake algorithms are discussed below:

- Face swap: In face-swapping methods, the identity of the person in the source video/image is replaced with that of the person in the target video. Here, the expression of the face in the source video or image is an important characteristic of this category of deepfakes. Methods such as FaceSwap, Deepfakes, and FaceShifter of FF++ dataset belong to this category. Recent methods of this category generate highly realistic deepfake videos.

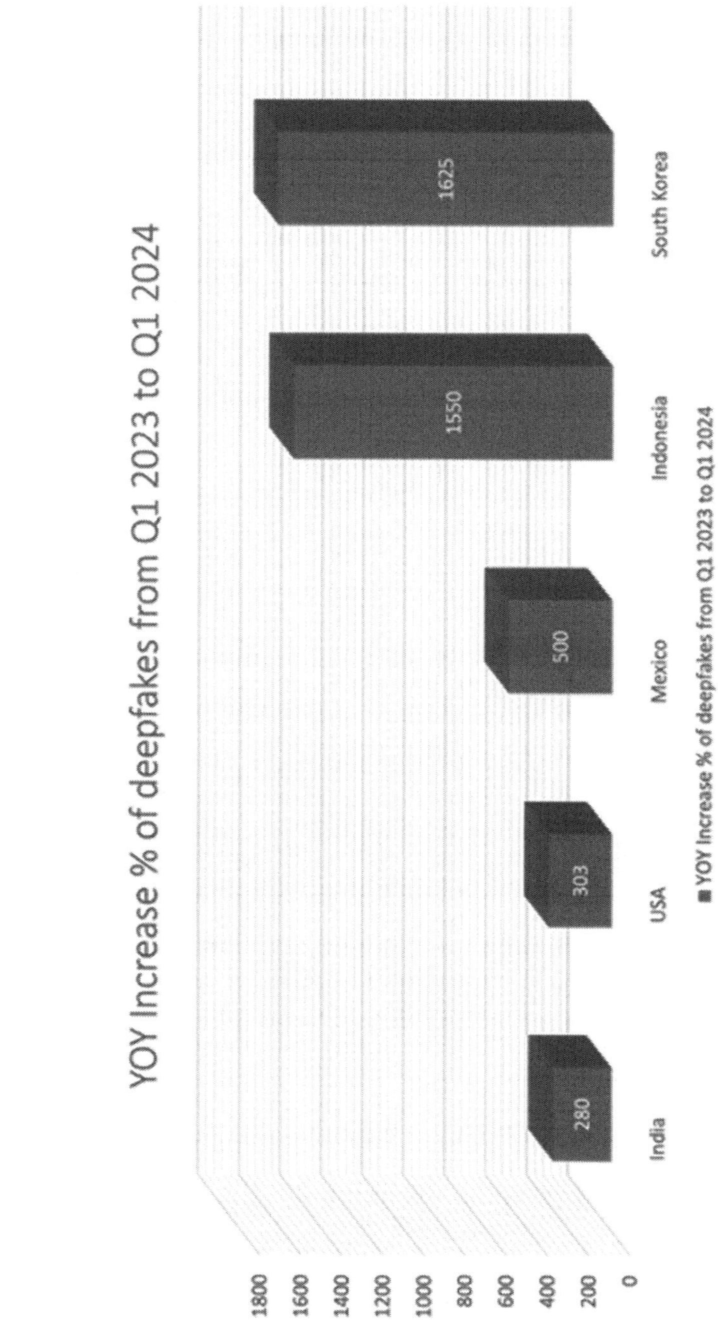

Figure 9.1 Country-wise Year on Year increase % of Deepfakes from Q1 2023 till Q2 2024 [2].

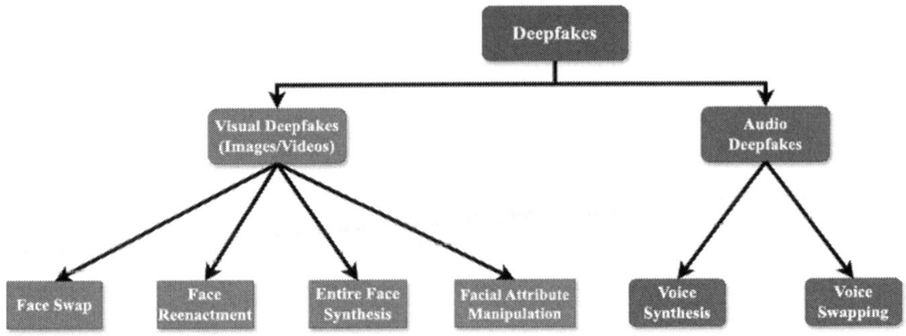

Figure 9.2 Types of deepfakes and deepfake generation methods.

- Face reenactment: In face reenactment methods, a portion of the face is modified to change the expression or say or do something that was never done or said. Although face reenactment methods often produce high-quality and realistic fake images/frames, they lack temporal consistency in the case of videos. Methods such as Face2Face of FaceForensics++ (FF++) [3] is an example face reenactment method. In this category, the aim of the deepfake generation methods is to reconstruct facial expressions in the source video or image of the face present in the target video while preserving the illumination, head pose, identity, and other parameters of the target video/image.
- Entire face synthesis: In entire face synthesis methods, a complete face, which is fake, is synthesized after training the generating model to learn the complex latent representations and features from the face dataset. Here, the synthesized face does not belong to any real person. ProGAN [4] belongs to this category of deepfake generation methods.
- Facial attributes manipulation: This category of deepfake generation methods includes methods which manipulate certain characteristics of the face, such as skin or eye or hair color, hairstyle, gender, age, etc. An example of this category of methods is StarGAN [5], which can perform image-to-image translation across multiple domains (or attributes of the face).

9.1.3 Deepfake generation process

Key technology used for creating deepfakes is generative adversarial network (GAN) [6] which was proposed in 2014. GAN as shown in Figure 9.3 consists of two networks which make it possible to generate new synthetic fake images. These comprising networks are (1) "generator", whose purpose

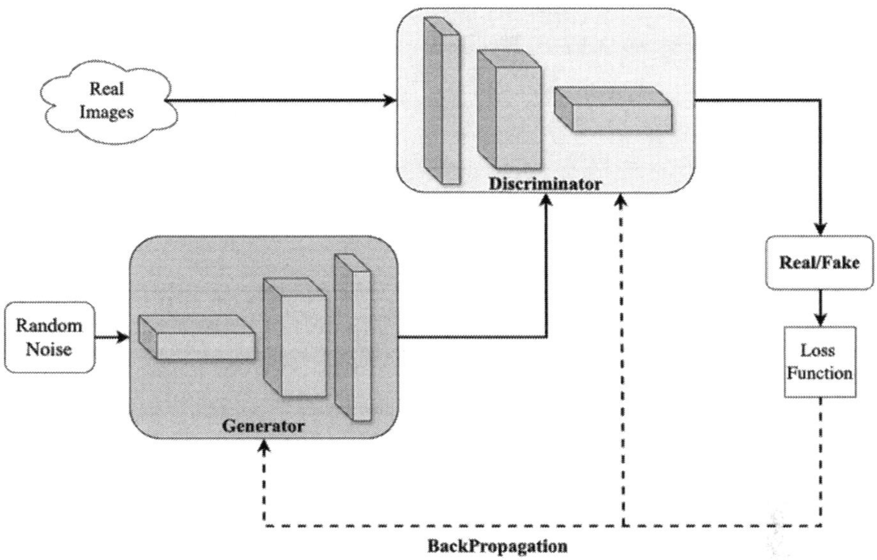

Figure 9.3 GAN architecture.

is to generate fake images, and (2) "discriminator", whose task is to classify if the image generated by the generator is from a dataset of real images.

The discriminator is trained like a binary classifier to maximize the probability of associating the correct label to both training samples as well as samples from the generator. The cost incurred by both networks is attempted to be minimized, and the back-propagation algorithm uses the derivatives of the discriminator's output with respect to the discriminator's input to train the generator to improve its output.

The first case of deepfake occurred in November 2017, when a Reddit user "deepfakes" published a pornographic video with the face of the porn actor replaced by the face of a celebrity using deep learning. Since then, there has been a rapid development in the field of deepfake such that celebrities, politicians, friends, and colleagues, and almost every person in society have been affected by it.

Since then, there has been an exponential rise in GAN methods, which are used in combination with other methods for generating deepfakes. ExprGAN [7] introduced facial expression manipulation with expression control ability. ProGAN [4] described a new methodology of GAN by progressively training generator and discriminator starting from low-resolution fake images and, as training progresses, adding new layers to synthesize finer quality images. StarGAN [5] proposed image-to-image translations on multiple domains using a single model. GANimation [8] allowed facial expression manipulation in the continuous manifold. StyleGAN [9] made it possible to synthesize highly varied and high-quality face images with scale control variation of synthesis of fake images. StyleGAN2 [10] is an improved version of StyleGAN. HomoInterpGAN [11] generates high-quality facial expression translation. AttGAN [12] provided realistic attribute manipulation.

Various other deepfake generation methods in the category of Face swap methods such as RSGAN [13], FSGAN [14], SimSwap [15], HiFiFace [16, 17], and face reenactment methods such as ReenactGAN [18, 19], ICface [20, 21] etc have also emerged.

Comparable to GANs, diffusion models [22] can synthesize sophisticated and good quality images [23].

9.1.4 Opportunities

- Marketing industry: Deepfakes provide new avenues for the marketing industry with the potential to revolutionize the industry by creating more personalized marketing and multilingual and cost-effective content creation. Under ethical and legal guidelines, digital marketing can use deepfakes to advertise a product using an individual's personal favorite celebrity using his/her mother tongue to increase customer engagement. Dalí Museum, Florida used deepfake Salvador Dalí to greet visitors [24]
- Entertainment industry: Deepfakes has various exciting, revolutionary, and desirable offerings to the entertainment industry. Deepfakes show the ability to create cost-effective, fast, and automatic solutions for re-ageing actors or replacing stunt doubles. Existing facial reageing and replacement methods are costly and time-consuming processes as they require manual frame-by-frame work, which often takes a lot of days even by skilled workers and requires costly CGI or Makeup work. With Deepfakes, facial reageing and replacement tasks can be made cost-effective and faster than existing methods. In 2022, researchers at Disney [25] presented a neural network-based production and fully automatic technique for reageing facial images in videos. It has the potential to make the arduous task of dubbing in different languages easier and cost-effective.
- Healthcare industry: Deepfake has the potential of helping people suffering from Alzheimer's disease to interact with people by creating their younger face, which can help patients to recall some memory [26]. Deepfake also has the potential to help and cure mental health patients by creating realistic simulations of people interacting with

them, providing an environment which they can find relatable and help create more interactive sessions between doctors and patients.

9.1.5 Threats

Due to the general availability of fake videos and images generating apps [27, 28], [29] and source code, various threats are posed by deepfakes such as:

1. Scammers using deepfakes to scam people by exploiting their socially available videos and images
2. Forging media and spoofing the identity of public figures or celebrities to damage their reputation as well as create chaos and unrest in society
3. Manipulating publicly available information present on the internet to influence the public in events such as elections in democracies to cause political unrest
4. Using deepfake for bypassing biometric security systems, identity theft, cyberbullying, or harming the reputation of individuals by creating fake pornographic material, thus posing a threat to both personal and social security.
5. Corporate sabotage using deepfake, which may result in a drastic drop in stock prices, malign the company's reputation and monetary losses.
6. Producing deepfake as false evidence in courts of law in the absence of a robust deepfake detection method presents a major threat which needs significant attention of law enforcement
7. People's natural inclination to believe what they see also poses a major threat resulting from deepfakes.

9.2 DATASETS

In this section, we discuss the various popular deepfake video datasets. Figure 9.4 and Table 9.1 show the various three generations of deepfake datasets [30, 31].

9.2.1 UADFV

This is one of the early datasets which contains 49 real videos obtained from YouTube and 49 fake videos generated using DNN model using FakeAPP [33].

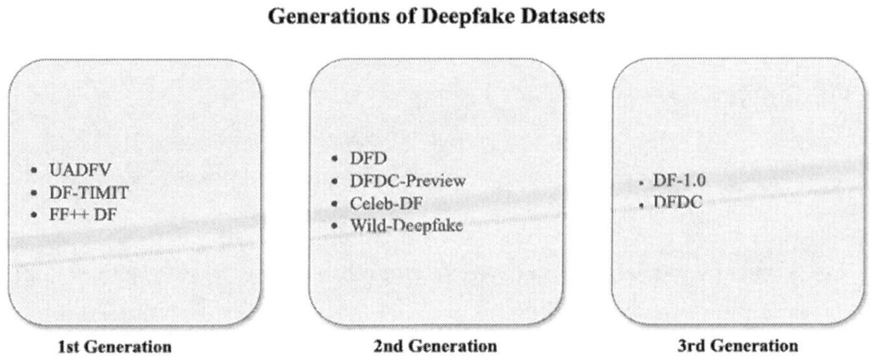

Figure 9.4 Generations of Deepfake datasets.

Table 9.1 List of Deepfake video datasets with their release years and real/fake sequences present in the dataset and source of videos of the dataset

Dataset	*Release Year*	*Real/Fake*	*Source*
UADFV	2018	49/49	YouTube
Deepfake-TIMIT	2018	-/620	VidTIMIT [32]
FaceForensics++	2019	1000/5000	YouTube
DFD	2019	363/3068	Actors
DFDC	2020	23654/104500	Actors
Celeb-DF (v2)	2020	590/5639	YouTube
WildDeepfake	2020	3805/3509	Internet
DF-1.0	2020	-/10,000	Actors
KoDF	2021	62,166/175,776	Actors

9.2.2 Deepfake-TIMIT

It is also one of the early-generation datasets that contain 690 Deepfake videos created using FaceSwap GAN on the VID-TIMIT dataset. It contains an equal number of high-quality videos (DF-TIMIT-HQ) and low-quality videos (DF-TIMIT-LQ) [34].

9.2.3 FaceForensics++ (FF++)

FaceForensics++ is a widely used large-scale dataset which originally contained a total of 5000 videos, of which 1000 videos are original videos obtained from YouTube and 4000 fake videos created using four different manipulation techniques, with 1000 fake videos created using each manipulation method. Later, a new manipulation technique, FaceShifter, was added to FaceForensics++ to add 1000 more manipulated videos, resulting in a total of 6000 videos. It is one of the benchmark datasets in which original videos have unoccluded frontal faces on which the following five manipulation methods have been applied [3]:

- Deepfakes [35]: It is a based method that is trained using pairs of subject images to compute a mapping between them and produces an image by face swapping the given face in the source image with the target face.
- FaceSwap [36]: It is a 3D face-swapping method that uses the traditional graphics pipeline and blending to create a forged image by swapping the face in the given image with that of the target.
- Face2Face [37]: It is a facial reenactment method that is based on computer graphics based approach. It uses expression manipulation of a target video from the source video while maintaining the facial identity of the target video.
- NeuralTextures [38]: It is a forgery method which learns a face neural texture from a target video and then uses it to create a realistic face reenactment while preserving the identity by only modifying the mouth region of the face.
- FaceShifter [39]: This method is added latest in the original FF++ dataset, and it is a two-stage face swapping method for producing high fidelity and occlusion aware face swapping by reconstructing attributes that are hard to replicate.

Face2Face and FaceSwap are computer graphics-based approaches, while Deepfakes and NeuralTextures are learning-based approaches. These videos are available in three types of compression quality: c23 quality videos (high-quality videos with low compression ratio), c40 quality videos (low-quality videos with high compression ratio), and raw videos.

9.2.4 DFD

DeepFake detection dataset is a dataset provided by Google and Jigsaw, which contains 3068 manipulated videos from 363 original videos obtained with the consent of actors of different genders, ages and ethnicities [40].

9.2.5 DFDC [31] and DFDC preview [41]

DeepFake Detection Challenge Dataset (DFDC) [31] is one of the largest datasets provided by Facebook which contains total 128,154 videos comprising of 104500 fake videos and 23654 real videos. These videos are created using 3426 paid actors of various ages, gender, skin-tone etc, and fake videos are generated using eight different Deepfake, GAN-based methods and non-learning-based methods. DFDC preview [41] comprises 4119 fake videos created and 1131 original videos with total videos of 5K videos of paid actors with fake videos created using two face manipulation methods.

9.2.6 Celeb-DF v2

Celeb-DF is one of the most challenging large-scale datasets, which contains 5639 deepfake videos generated from 590 original videos. The source of these videos is publicly available YouTube videos of 59 celebrities of diverse ages, ethnicity, and gender. It is one of the most difficult data set for existing Deepfake detection algorithms [30].

9.2.7 WildDeepfake

WildDeepfake is one the recent datasets consisting of 707 total videos sourced from the internet providing 3805 real face sequences and 3509 fake face sequences representing a wide variety of wild deepfakes present in the internet [42].

9.2.8 DeeperForensics-1.0 (DF-1.0)

It is one of the large datasets containing a total of 60000 videos with real and fake videos in the ratio of 5:1. Videos of 100 paid actors are used, which are created using a controlled indoor environment, and fake videos are created by applying seven different types of perturbations at five different intensity levels [43].

9.2.9 KoDF

It is the largest dataset, containing a total of 237,942 video clips of 403 subjects comprising 175,776 fake videos and 62,166 real videos. Fake videos are synthesized using six different varieties of face-swap and face-reenactment methods [44].

9.3 VISUAL DEEPFAKE DETECTION METHODS

9.3.1 Stages used in Deepfake video detection studies

Figure 9.5 shows the overall steps used in the deepfake video detection studies. These steps are summarized below:

- The Initial phase consists of the creation of a dataset by either collecting and organizing deepfake videos or using available datasets.

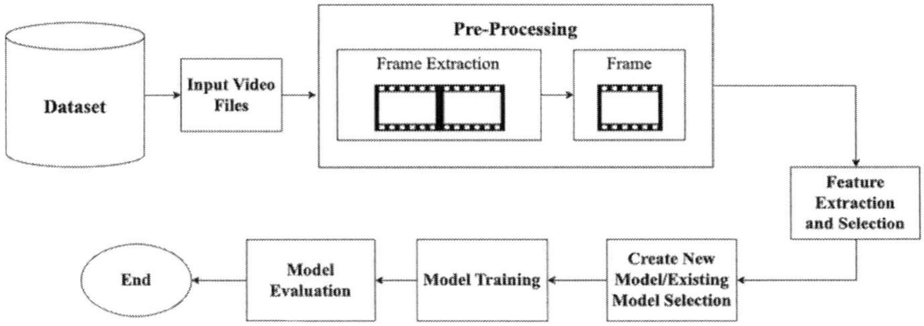

Figure 9.5 Steps used in Deepfake detection.

- From the dataset, videos are preprocessed by extracting the video frames and cropping and resizing as needed is performed on the frames to prepare for improved feature selection.
- Feature extraction is performed and useful features are selected to be used for detection
- The model to be used for detection is then either a newly created or a suitable available model.
- The selected model is then trained on the dataset to learn key features for distinguishing real and fake videos.
- Finally model evaluation is done using various performance metrics to assess the performance of the model.

9.3.2 Classification of detection methods

- Biological/Physiological signal based methods: use biological/physiological signals such as abnormalities in eye blinking, heart rate, skin color, etc., for detection of deepfakes.
- Statistical feature-based methods: employ statistical measures such as PRNU, cross-correlation scores, etc, for forgery identification.
- Machine learning based methods: work by defining features and the feature vector is fed to machine learning algorithms such as support vector machine, logistic regression, random forest, etc to train the classifier for detection of deepfakes.
- Frequency domain feature-based methods: are based on the identification of forgery clues in the frequency domain and help in the discovery of latent features which provide crucial clues for deepfake detection.
- Temporal domain feature-based methods: use temporal inconsistencies in video frames for deepfake detection.

- Deep learning-based methods: are widely used deepfake detection methods as they can extract complex latent patterns and learn features to provide high detection performance using deep learning models such as CNN, RNN, etc.
- Transformer-based methods: use transformers such as Vision Transformer (ViT) [45] for improved generalized deepfake detection by leveraging the self-attention mechanism to capture complex spatial and temporal dependencies within videos.

9.3.3 Early Deepfake detection methods

Early deepfake detection methods employed hand-crafted artefacts or CNN-based feature extraction methods to extract the visual features, which are different for manipulated and real images, and then train classifiers on the visual features to detect deepfake images.

1. *Biological signal based methods:* Early hand-crafted artefact methods included physiological or biological signals-based deepfake detection methods that use physiological/biological signs such as eye blinking or heart rate as signs for classifying videos as either fake or real.

   Li et al. [46] used eye blinking abnormalities in fake videos to detect deepfake by using Long term Recurrent CNN (LRCN) model to capture eye states. Each frame is cropped to a rectangular shape around the eye area after face alignment and passed to the LRCN model to learn discriminative temporal features, predict the eye states, and calculate blink frequency. The model shows good performance in detecting deepfake videos on the evaluation of self-made datasets.

   Tackhyun et al. [47] proposed an integrity verification algorithm that uses an eye-blinking pattern to classify a video as fake or real. The authors established an initial DB containing natural values of various factors that affect the eye blinking pattern such as age, gender, time of the day, and the activity of the subject in the video and corresponding information like blink count, period, elapsed time, etc. The proposed algorithm is used to compare the blinking information from the database with that extracted from the video to classify the video as fake or real. The method is evaluated on a self-made dataset of deepfake videos and achieves an accuracy of little over 87%.

   FakeCatcher [48], proposed by Aybars et al., is a biological signals-based deepfake detection model for fake portrait videos and uses color-based remote photoplethysmography (rPPG [49]) to extract heart rate signals by observing subtle changes in the skin color. The

PPG is a technique that can detect changes in skin color resulting from blood circulation. The model is evaluated using datasets FF++, UADFV, FF, and Celeb-DF in addition to their own dataset and achieving accuracies in the range 91%–96% for different datasets. Similarly, Hernandez-Ortega et al. [50] is also a biological signals-based model that uses a convolutional attention network (CAN) trained for heart rate estimation using rPPG to extract spatial and temporal information from video frames and analyze both the extracted information to detect fake videos. This model is evaluated on Celeb-DF and DFDC preview datasets and demonstrated AUC performance of 99.9 % and 98.2% on the datasets, respectively. However, biological/physiological signals-based methods are no longer efficient as the latest deepfake generation algorithms are able to reconstruct such signals with more efficiency and with good enough quality that such methods cannot efficiently detect abnormalities associated with biological or physiological signals.

2. *Early CNN-based methods:* Early learning-based approaches used CNN-based methods such as [51], which is based on the computation of co-occurrence matrix on the RGB channels of an image and used the computed vector as input to deep CNN model to detect deepfakes.

9.3.4 Statistical feature-based methods

Alongside these methods, statistics-based deepfake fingerprint identification methods mentioned in [52] and [53] used photo response nonuniformity (PRNU), noise pattern resulting by light-sensitive sensors of digital cameras for detection of deepfakes.

Luca et al. [54] proposed deepfake detection approach that is based on an expectation and maximization algorithm to create a mathematical model to represent the local correlations between each pixel and its neighborhood and capture this relationship in the form of a vector called convolutional trace (CT) that characterizes the traces left by GAN process used for creating deepfake images. The method extracts CT and uses it as a feature vector as input to naïve classifier to discriminate between real and fake images. Experimentations were conducted on images generated by 10 different GANs which involved with and without face manipulation and using different classifiers such as SVM, LDA, K-NN etc. The Random forest classifier provided the best classification result with an overall accuracy of over 98%, and when tested on deepfakes generated by FACEAPP, a test accuracy of 93% was achieved. However, accuracy decreases to as low as 56.96% on the application of different postprocessing techniques such as blurring, rotation, compression, etc.

9.3.5 Frequency domain feature-based methods

Frequency in Face Forgery Network (F^3 net) proposed by Qian et al. [55] used frequency-aware forgery clues for face forgery detection. The proposed method consists of two branches: (1) frequency aware decomposition branch, which partitions input image in the frequency domain according to learnable filters, and then decomposed frequency components are transformed back into spatial domain image components, which are stacked together and fed to CNN to fine forgery clues. (2) The local frequency statistics branch mines discrepancies in local frequency statistics between real and fake images of faces by employing sliding window DCT on the input image and calculating the mean frequency values within a set of learnable frequency bands. The two branches are then fused together using a cross-attention fusion module. The features of the two branches are then augmented using cross attention fusion module for feature interaction between the two branches and FF++ dataset is used for method evaluation and results of which show good performance in different qualities of images of FF++ with a minimum of 93.2% accuracy and 0.958 AUC score for low quality and a maximum of 99.99% accuracy and 0.9999 AUC score for raw FF++ videos.

Another frequency-based deepfake detection method, hierarchical frequency-assisted interactive networks (HFI-Net) proposed by Changtao et al. [56] is dual branch network that explores mining face forgery clues of deepfakes in the frequency domain. It consists of two branches: (1) transformer branch, which consists of four stages of transformer blocks (each stage has three ViT transformer blocks), captures Global forgery clues (2) CNN branch, for extracting local forgery artefacts, also has four stages as transformer branch with each stage containing three separable convolution blocks and a residual connection between input and output of the stage. At each stage of HFI-Net a global–local interaction (GLI) module fuses, global and local information and allows interaction between the two branches of the network and combines them to provide complementary features and improve generalization ability. Intradataset experimentation is done on FaceForensics++ (FF++) dataset, and cross-dataset evaluation is done on Celeb-DF, TIMIT, DFDC, and UADFV datasets. and also use GID-DF and GID-FF protocols [57] for unseen manipulations experimentation and DeeperForensics-1.0 for unseen perturbations experiments. The method shows impressive performance in different experiments.

9.3.6 Temporal domain feature-based methods

The recurrent neural network method proposed by Xi et al. [58] is based on extracting spatial, steganalysis, and temporal features. For spatial features, XceptionNet is used, and for extracting Steganalysis features, constrained convolutional filter is used on simplified XceptionNet. Both of these features are fused together and temporal features are extracted from the fused features of the consecutive frames using LSTM. The proposed method is evaluated using FF++ and GAN-based Deepfakes [59] dataset. The method achieves accuracy % of 98.57% (FF++ HQ) and 90.11% (FF++ LQ) in the intradataset evaluation, and cross-dataset evaluation on GAN-based Deepfakes achieves accuracy comparatively lower than that of intra-dataset evaluation.

Xi et al. [60] proposed a deepfake video detection method that uses both spatial and noise features of the spatial and temporal domain of video frame to detect deepfake. Of the k-selected video frames, each cropped facial regions are sent to two streams: RGB stream and SRM (Spatial Rich Model) stream. RGB stream derives the RGB features and SRM stream is used for extracting noise features by applying learnable SRM filters whose values are defined manually. The fusion of both segment-based and stream-based predictions provides the final detection result. Both streams use Xception as a backbone network, and the method is evaluated on the FF++ dataset and achieves the performance of 90.36% accuracy on evaluation.

9.3.7 Deep learning-based methods

Yuval et al. [61] presented a method which aimed at identifying deepfake videos based on comparing inner and outer context regions, and the method is based on the assumption that the manipulation method only manipulates the inner part of the face and the outer context regions such as hairs, neck, ears etc are not easy to regenerate by the manipulation methods. During the preprocessing step, the face region is detected and cropped, and the inner face region and context region are obtained and fed into different Xception [62]-based networks. E_f learns on the facial region, and E_c learns on the context region. Extracted vectors are then the difference of the two extracted vectors is computed and used as discrepancy vector which is concatenated with vectors obtained from the other two networks which map face regions from face swap and face reenactment methods. The concatenated vector is then passed to the final classifier for the final detection task. The method uses multiple datasets such as VGGFace2, Labeled Faces in the Wild (LFW), FaceForensics++, DFDC, Celeb-DFDC, and FaceForensics++ datasets for training and testing. This method achieves AUC % performance of as low as 66% (Celeb-DF v2) and as high as 99.7% (FF-DF) in images created by face swapping.

Framework for mining generalized features for the detection of AI-manipulated faces proposed by Yang et al. [63] focused on mining intrinsic features in the form of channel difference image (CDI) and spectrum image (SI), which are indispensable characteristics of the camera imaging process and AI manipulation process. The framework uses the Octave convolution for mining the features as both the features are related to the frequency domain and uses an attention-based fusion module to effectively combine the extracted information and adopt a domain alignment module to improve generalizability. The dataset used for experiments is constructed by four categories (face entire synthesis, facial expression manipulation, facial attribute manipulation, and identity manipulation), and the dataset contains 10 state-of-the-art and popular manipulation techniques such as StyleGAN, StarGAN, FF++ (Deepfakes), DFDC, etc. The model outperforms other existing methods in both the same dataset analysis as well as cross dataset analysis and demonstrates good performance for different postprocessing methods such as compression, resizing, median filtering, noise addition, etc. For unseen categories, the method achieves an accuracy performance of 86.79% for entire synthesis, 89.26% for expression manipulation, 89.74% for attribute manipulation, and 81.92% for identity manipulation.

Based on the observation that deepfake generation methods generate inconsistent lighting, sharpness, noise, blur, etc, which results in loss of depth information in fake images as compared to real images, Liang et al. [64] proposed a depth map prediction network and depth map guided triplet network. depth map prediction network evaluates depth information and generates a depth map. The triplet feature extraction network has an input of the original face, fake face and target face each concatenated by a depth map extracted from the depth prediction network which forms triplet input and consists of three swin transformer-based feature extractors which are trained on triplet and use the triplet loss to constraint the feature distance between two real images as close as possible and the distance between real and fake images as far as possible. Finally, these extracted features are combined and fed as input into the classifier for classification. Biwi Kinect Head Pose Database, FF++, and Celeb-DF dataset are used for training and testing of the proposed method and achieve a minimum of 91.51% accuracy (in case of low-quality FF++ dataset) and AUC % of 72.30 in case of Celeb-DF dataset in cross-dataset evaluation scenario.

Liang et al. [65] proposed a framework called masked relation learning for video-based Deepfake detection which consisted of two components: (1) SpatioTemporal Attention (STA) module which produces attention maps to extract features from various facial regions and (2) masked relation learner (MRL) module for learning the irregularity in the relation between facial regions from deepfakes. Firstly, a graph is constructed in which vertices denote features of facial regions and the edge represents feature correlation between two regions. In the training procedure, MRL masks the partial edges to reduce the redundancy of relations and learn the critical structure of faces and uses the temporal convolution network (TGCN) to capture the irregularities of relations to expose the deepfakes. FF++, Celeb-DF, and DFDC are used as datasets for evaluation of the proposed method and achieve competitive performance of AUC% performance of 96.18% in FF++ (c40) dataset, 83.58% in Celeb-DF, and 71.53% in DFDC dataset.

Multiclassification framework FCD-Net of deepfake images to detect multiple types of deepfakes proposed by Ruidong et al. [66] consists of three modules to detect deepfake images and to classify them. These three modules are: (1) Face Synaptic Saliency Module (FSS) module which consists of three parallel branches of depth separable convolution layers is used for extracting better face features than standard convolution layers. (2) Contour detail feature extraction (CDFE) Module is used for extracting key discriminating facial features and reduce impact of background noise. (3) Distinguishing feature fusion (DFF) Module is an improved pooling layer used before fully connected layer to extract multiscale information by concatenating multiple max pooling layers of different kernel sizes. In addition, they proposed HDFD dataset of homologous deepfake images which consists of four categories of images: real face images (obtained from FFHQ dataset), face swap synthesis, entire face synthesis, and facial attribute manipulation, with each category containing 8000 images where each manipulated image is constructed from the real ones through StyleGAN, FaceAPP, and Deepfakes(FaceSwap). The proposed method is then trained and tested on the HDFD dataset with AUC % performance reaching nearly 100%.

1. *Transformer based detection methods:* Han et al. [67] proposed a method to detect deepfakes by learning distinctive features of intraconsistency within classes and inter-diversity between classes. Firstly, a Vision Transformer (ViT) [45]-based encoder is trained by using patch-wise masking and prediction scheme where the encoder is trained to predict masked image patches, which help in mining hidden features in an image without using deepfake labels in a self-supervised learning way. For better generalizability and improved latent discriminative feature extraction, the pretrained encoder is fine-tuned with labelled data for deepfake detection. To develop a more general and robust detection method, a Discrete Wavelet Transformation (DWT)-based Adjustable Forgery Synthesizer (AFS) module is introduced, which augments the data to generate diverse quality and diverse similarities between images to facilitate better representation learning. Datasets used for evaluation are: FF++, CelebDF, DFDC, UADFV, DF-TIMIT, DFR, and WildDeepfake and demonstrate competitive performance. The method shows the highest AUC % of 99.55% in the case of the FF++ (c23) dataset and the lowest of 81.44% in the case of the WildDeepfake dataset (model trained on FF++ DF (c23)).

Tianyi et al. [68] proposed a keyframe-based deep convolutional transformer model, which jointly leveraged convolutional pooling and reattention approaches to study the discriminative image features. Local features of the image are extracted using a stack of CNN layers, and global features are

Table 9.2 Summary of performance of detection methods in intradataset scenario

Year	*Reference*	*Dataset*	*Performance*
2018	[46]	Self-generated Dataset	99.0 (AUC%)
2020	[47]	Self-generated Dataset	87.5 (ACC%)
2020	[48]	Self-generated Dataset	91.07 (ACC%)
2020	[55]	FF++ (c23)	87.01 (FF++ c23), 67.35 (FF++ c40) (ACC %)
		FF++ (c40)	75.03 (FF++ c23), 75.50 (FF++ c40) (ACC %)
2020	[58]	FF++	98.57 (FF++ c23), 90.11 (FF++ c40) (ACC%)
2021	[60]	FF++	90.36 (ACC %)
2022	[56]	FF++	94.19 (FF++ c40), 99.58 (FF++ c23) (AUC%)
2022	[61]	FF++	75.00 (ACC %)
2023	[64]	FF++	99.80 (AUC%)
2023	[65]	FF++	98.27 (FF++ c23),96.18 (FF++ c40) (AUC%)
2023	[66]	Self-generated Dataset	99.99 (AUC%)
2023	[67]	FF++ DF (c23)	99.55 (FF++ c23), 98.03 (FF++ c40) (AUC%)
		FF++ DF (c40)	99.50 (FF++ c23), 99.32 (FF++ c40) (AUC%)
2023	[68]	FF++	97.66 (AUC%)

analyzed using ViT together with convolution pooling and reattention technique on keyframes or I-frames, which contain complete information than using P-frames and B-frames which suffer from information loss as a result of compression. Datasets used for evaluation are FF++, Celeb-DF, DFDC, and DF-1.0 and results show that the method has good performance in the same dataset scenario of however cross-dataset performance is relatively lower.

Table 9.2 provides a summary of the methods in an intradataset scenario where training and testing are done on the same dataset. Table 9.3 summarizes the performance of the deepfake detection methods in cross-dataset scenarios where training and testing are done on different datasets.

To assess the performance of various methods, various metrics used for evaluation of deepfake detection methods are shown in Table 9.4. It has been observed that the most often used metrics are accuracy, receiver operating characteristic (ROC) curve, and AUC. While error rate, recall, and F1-score are comparatively not popular in deepfake detection methods, accuracy and AUC are the most popular metrics used for performance evaluation.

9.4 EMERGING CHALLENGES AND OPEN ISSUES

With the rise of deepfakes, new challenges are emerging in the way of deepfake detection methods. These challenges can be summarized as follows:

- Decreased cost of creating doctored media as deep learning has made it easy for anyone to perform manipulation in less time and effort

Table 9.3 Summary of performance of detection methods in cross-dataset scenario

Year	*Reference*	*Training Dataset*	*Testing Dataset*	*Performance*
2020	[48]	Celeb-DF	FF++ Celeb-DF	83.10 (ACC%)
		FF++		86.48 (ACC%)
2020	[58]	FF++	Deepfake-TIMIT (HQ)	84.63, 83.78 (ACC%)
		FF++	Deepfake-TIMIT (LQ)	97.89, 95.01 (ACC%)
2022	[56]	FF++ (c40)	Celeb-df	82.37 (AUC%)
		FF++ (c23)	Celeb-df	83.28 (AUC%)
2022	[61]	FF++	Celeb-DF	66 (AUC%)
2023	[64]	FF++	Celeb-DF	72.30 (AUC %)
2023	[65]	FF++ c40	Celeb-DF	83.58 (AUC %)
		FF++ c40	DFDC	71.53 (AUC %)
2023	[66]	Self-generated dataset	Self-collected dataset	82.33 (AUC %)
2023	[67]	FF++ DF (c23)	Celeb-DF	86.24 (AUC%)
		FF++ DF (c40)	Celeb-DF	86.84 (AUC%)
2023	[68]	FF++	Celeb-DF	72.43 (AUC%)
		FF++	DFDC	73.68 (AUC%)

Table 9.4 Performance metrics used by various studies for Deepfake detection

Metrics	*Definition*
Accuracy (ACC)	Accuracy is the measure of total number of correct predictions out of all predictions. ACC = (TN + TP)/ (TP+FP+TN+FN)
ROC curve	Receiver operating characteristic curve or ROC curve is a graphical plot to represent performance graphically by showcasing trade-off between true positive rate (TPR) and false positive rate (FPR) at different thresholds and separates signal from noise
Area Under Curve (AUC)	The AUC represents the area under ROC curve. The greater the AUC, the better the performance of the model. If AUC is equal to 1, the classifier is able to perfectly distinguish between all the Positive and Negative instances.
Precision	Precision refers to the proportion of correct positive predictions by the model out of all positive predictions. Precision = TP/ (TP+FP)
Recall	Recall, also known as sensitivity, provides a measure of the proportion of all correct positive predictions out of all the actual positive instances. Recall = TP / (TP + FN)
F1 score	F1 scores measure the harmonic mean of precision and recall to provide a comprehensive evaluation of the model. F1 = 2 * Recall * Precision / (Recall + Precision)
log loss	log loss assess the performance of a classification model and important metrics used for classification
Error Rate	Error rate assess the performance by providing the number of instances the classification model predicts wrongly. Error Rate = (FP + FN)/ (TP + TN + FP + FN)
MCC	Matthews correlation coefficient or MCC provides a single value summarizing the confusion matrix

unlike earlier times when an individual needed to be skilled in some image processing tool to create realistic morphed images
- Continuous evolution of powerful deepfake generation methods and new methods such as variational autoencoders (VAEs) [69] and diffusion models [22] that create very realistic doctored media
- Increasing computing capabilities of devices and easy availability of powerful hardware like GPUs have increased generation of hyper-realistic deepfakes
- Public availability of the source code and apps for generating deepfakes such that anyone can generate deepfakes easily with just access to the internet

9.5 METRICS

9.5.1 Open issues

In addition to the emerging challenges, there are several open issues which need the attention of future researches in the field of deepfake detection.

- High computational cost: Most of the existing methods require high computations and significant time to perform deepfake detection. This prevents deepfake detection on real-time data. This creates a gap that needs to be addressed by future researchers. to develop detection methods with lesser computations and time consumption.
- Lack of good generalized methods: An analysis of various existing methods shows the need to improve the generalization ability of detection methods as well as the need for performance improvement in postprocessed videos and requires immediate attention of researchers of both industry and academics for developing tools and techniques that can detect the deepfakes with high efficiency and robustness and reliability to create a check on misinformation by providing public in general the ways to be able to distinguish between fake and real.
- Need for balanced quality dataset: Most of the existing deepfake datasets are imbalanced, and as new variants of deepfakes are emerging and evolving continuously, there is a need for a balanced dataset that contains a collection of real and fake data created using diversified manipulation methods and perturbation methods.

9.6 CONCLUSION

In this survey, we presented a detailed survey of deepfakes from their origin and generation process to various deepfake detection methods developed over the years. We found that the generation process of deepfakes is continuously evolving, and there has been a steep rise in deepfakes circulation on the internet over the last few years. FF++ is the most popular deepfake dataset that has been used in the majority of deepfake detection methods, with Accuracy being the most popular evaluation metric. We also observed that CNN-based methods have been the most widely used techniques used for deepfake detection.

With the rapid rise in the circulation of deepfakes on the internet and the continuous evolution of the deepfake generation process, deepfake detection methods still face a wide variety of challenges which need increased attention from researchers. This article covers a wide range of topics related to deepfakes such as categories of deepfakes, datasets available on the internet for deepfake detection methods' training and testing, various existing deepfake detection methods as well as threats and challenges related to deepfakes etc so that this article can serve as a valuable resource and hope will help the researchers in developing effective deepfake detection methods.

REFERENCES

[1] " 'Deepfakes' ranked as most serious AI crime threat," ScienceDaily, Aug. 20, 2020. Available: www.sciencedaily.com/releases/2020/08/200804085908.htm

[2] Sumsub, "Deepfake cases surge in countries holding 2024 elections, Sumsub research shows," Sumsub, Jun. 5, 2024. [Online]. Available: https://sumsub.com/newsroom/deepfake-cases-surge-in-countries-holding-2024-elections-sumsub-research-shows/

[3] A. Rossler, D. Cozzolino, L. Verdoliva, C. Riess, J. Thies, and M. Nießner, "Faceforensics++: Learning to detect manipulated facial images." in *Proceedings of the IEEE/CVF International Conference on Computer Vision*, pp. 1–11, 2019.

[4] T. Karras, T. Aila, S. Laine, and J. Lehtinen, "Progressive growing of gans for improved quality, stability, and variation," *arXiv preprint arXiv:1710.10196.*

[5] Y. Choi, M. Choi, M. Kim, J. Ha, S. Kim, and J. Choo, "Stargan: Unified generative adversarial networks for multi-domain image-to-image translation," in *2018 IEEE/CVF Conference on Computer Vision and Pattern Recognition (CVPR)*, (Los Alamitos, CA, USA), pp. 8789–8797, IEEE Computer Society, Jun 2018.

[6] I. Goodfellow, J. Pouget-Abadie, M. Mirza, B. Xu, D. Warde-Farley, S. Ozair, A. Courville, and Y. Bengio, "Generative adversarial networks," *Commun. ACM*, vol. 63, p. 139–144, Oct 2020.

[7] H. Ding, K. Sricharan, and R. Chellappa, "Exprgan: Facial expression editing with controllable expression intensity," *CoRR*, vol. abs/1709.03842, 2017.

[8] A. Pumarola, A. Agudo, A. M. Martinez, A. Sanfeliu, and F. Moreno-Noguer, "Ganimation: Anatomically-aware facial animation from a single image," in *Computer Vision – ECCV 2018* (V. Ferrari, M. Hebert,

C. Sminchisescu, and Y. Weiss, eds.), (Cham), pp. 835–851, Springer International Publishing, 2018.

[9] T. Karras, S. Laine, and T. Aila, "A style-based generator architecture for generative adversarial networks," *CoRR*, vol. abs/1812.04948, 2018.

[10] T. Karras, S. Laine, M. Aittala, J. Hellsten, J. Lehtinen, and T. Aila, "Analyzing and improving the image quality of stylegan," *CoRR*, vol. abs/1912.04958, 2019.

[11] Y.-C. Chen, X. Xu, Z. Tian, and J. Jia, "Homomorphic latent space interpolation for unpaired image-to-image translation," in *2019 IEEE/CVF Conference on Computer Vision and Pattern Recognition (CVPR)*, pp. 2403–2411, 2019.

[12] Z. He, W. Zuo, M. Kan, S. Shan, and X. Chen, "Attgan: Facial attribute editing by only changing what you want," *IEEE Transactions on Image Processing*, vol. 28, no. 11, pp. 5464–5478, 2019.

[13] R. Natsume, T. Yatagawa, and S. Morishima, "Rsgan: face swapping and editing using face and hair representation in latent spaces," in *ACM SIGGRAPH 2018 Posters*, SIGGRAPH '18, (New York, NY, USA), Association for Computing Machinery, 2018.

[14] Y. Nirkin, Y. Keller, and T. Hassner, "Fsgan: Subject agnostic face swapping and reenactment," in *2019 IEEE/CVF International Conference on Computer Vision (ICCV)*, (Los Alamitos, CA), pp. 7183–7192, IEEE Computer Society, Nov 2019.

[15] R. Chen, X. Chen, B. Ni, and Y. Ge, "Simswap: An efficient framework for high fidelity face swapping," in *Proceedings of the 28th ACM International Conference on Multimedia*, MM '20, (New York, NY), p. 2003–2011, Association for Computing Machinery, 2020.

[16] Y. Wang, X. Chen, J. Zhu, W. Chu, Y. Tai, C. Wang, J. Li, Y. Wu, F. Huang, and R. Ji, "Hififace: 3d shape and semantic prior guided high fidelity face swapping," *CoRR*, vol. abs/2106.09965, 2021.

[17] Y. Zhu, Q. Li, J. Wang, C. Xu, and Z. Sun, "One shot face swapping on megapixels," in *2021 IEEE/CVF Conference on Computer Vision and Pattern Recognition (CVPR)*, (Los Alamitos, CA), pp. 4832–4842, IEEE Computer Society, Jun 2021.

[18] W. Wu, Y. Zhang, C. Li, C. Qian, and C. C. Loy, "Reenactgan: Learning to reenact faces via boundary transfer," in *Computer Vision – ECCV 2018* (V. Ferrari, M. Hebert, C. Sminchisescu, and Y. Weiss, eds.), (Cham), pp. 622–638, Springer International Publishing, 2018.

[19] Y. Zhang, S. Zhang, Y. He, C. Li, C. C. Loy, and Z. Liu, "One-shot face reenactment," *CoRR*, vol. abs/1908.03251, 2019.

[20] S. Tripathy, J. Kannala, and E. Rahtu, "Icface: Interpretable and controllable face reenactment using gans," in *2020 IEEE Winter Conference on Applications of Computer Vision (WACV)*, pp. 3374–3383, 2020.

[21] W. Kang, G. Lee, H. I. Koo, and N. I. Cho, "One-shot face reenactment on megapixels," *ArXiv*, vol. abs/2205.13368, 2022.

[22] J. Ho, A. Jain, and P. Abbeel, "Denoising diffusion probabilistic models," in *Advances in Neural Information Processing Systems* (H. Larochelle, M. Ranzato, R. Hadsell, M. Balcan, and H. Lin, eds.), vol. 33, pp. 6840–6851, Curran Associates, Inc., 2020.

[23] P. Dhariwal and A. Nichol, "Diffusion models beat gans on image synthesis," *CoRR*, vol. abs/2105.05233, 2021.
[24] R. S. Aouf and R. S. Aouf, "Museum creates deepfake Salvador Dalí to greet visitors," Dezeen, Nov. 26, 2020. [Online]. Available: www.dezeen.com/2019/05/24/salvador-dali-deepfake-dali-musuem-florida/
[25] G. Zoss, P. Chandran, E. Sifakis, M. Gross, P. Gotardo, and D. Bradley, "Production-ready face re-aging for visual effects," *ACM Transactions on Graphics*, vol. 41, pp. 1–12, 2022.
[26] M. Westerlund, "The emergence of deepfake technology: A review," *Technology Innovation Management Review*, vol. 9, pp. 39–52, Nov 2019.
[27] L. Guilloux, "FakeApp 2.2 – Download for PC Free," Malavida, Mar. 06, 2019. www.malavida.com/en/soft/fakeapp/
[28] "FaceApp: Face Editor." www.faceapp.com/
[29] "Home." www.wombo.ai/
[30] Y. Li, X. Yang, P. Sun, H. Qi, and S. Lyu, "Celeb-df: A large-scale challenging dataset for deepfake forensics," in *2020 IEEE/CVF Conference on Computer Vision and Pattern Recognition (CVPR)*, pp. 3204–3213, 2020.
[31] B. Dolhansky, J. Bitton, B. Pflaum, J. Lu, R. Howes, M. Wang, and C. Canton-Ferrer, "The deepfake detection challenge dataset," *CoRR*, vol. abs/2006.07397, 2020.
[32] "Conrad Sanderson – VidTIMIT dataset." https://conradsanderson.id.au/vidtimit/
[33] X. Yang, Y. Li, and S. Lyu, "Exposing deep fakes using inconsistent head poses." in ICASSP 2019-2019 IEEE International Conference on Acoustics, Speech and Signal Processing (ICASSP), pp. 8261–8265. IEEE, 2019.
[34] P. Korshunov and S. Marcel, "Deepfakes: a new threat to face recognition? assessment and detection." https://arxiv.org/abs/1812.08685, 2018.
[35] Deepfakes, "GitHub – deepfakes/faceswap: Deepfakes Software For All," GitHub. https://github.com/deepfakes/faceswap
[36] MarekKowalski, "GitHub – MarekKowalski/FaceSwap: 3D face swapping implemented in Python," GitHub. https://github.com/MarekKowalski/FaceSwap/
[37] J. Thies, M. Zollhofer, M. Stamminger, C. Theobalt, and M. Nießner, "Face2face: Real-time face capture and reenactment of rgb videos," in *2016 IEEE Conference on Computer Vision and Pattern Recognition (CVPR)*, pp. 2387–2395, 2016.
[38] J. Thies, M. Zollhofer, and M. Nießner, "Deferred neural rendering: image synthesis using neural textures," *ACM Transactions on Graphics*, vol. 38, pp. 1–12, Jul 2019.
[39] L. Li, J. Bao, H. Yang, D. Chen, and F. Wen, "Advancing high fidelity identity swapping for forgery detection," in *Proceedings of the IEEE/CVF Conference on Computer Vision and Pattern Recognition*, pp. 5074–5083, 2020.
[40] "Contributing data to deepfake detection research." https://research.google/blog/contributing-data-to-deepfake-detection-research
[41] B. Dolhansky, R. Howes, B. Pflaum, N. Baram, and C. Canton-Ferrer, "The deepfake detection challenge (DFDC) preview dataset," *CoRR*, vol. abs/1910.08854, 2019.

[42] B. Zi, M. Chang, J. Chen, X. Ma, and Y.-G. Jiang, "Wilddeepfake: A challenging real-world dataset for deepfake detection," in *Proceedings of the 28th ACM International Conference on Multimedia*, MM '20, (New York, NY), p. 2382–2390, Association for Computing Machinery, 2020.

[43] L. Jiang, W. Wu, R. Li, C. Qian, and C. C. Loy, "Deeperforensics-1.0: A large-scale dataset for real-world face forgery detection," *CoRR*, vol. abs/2001.03024, 2020.

[44] P. Kwon, J. You, G. Nam, S. Park, and G. Chae, "Kodf: A large-scale korean deepfake detection dataset," in *Proceedings of the IEEE/CVF International Conference on Computer Vision (ICCV)*, pp. 10744–10753, October 2021.

[45] A. Dosovitskiy, L. Beyer, A. Kolesnikov, D. Weissenborn, X. Zhai, T. Unterthiner, M. Dehghani, M. Minderer, G. Heigold, S. Gelly, J. Uszkoreit, and N. Houlsby, "An image is worth 16x16 words: Transformers for image recognition at scale," *CoRR*, vol. abs/2010.11929, 2020.

[46] Y. Li, M. Chang, and S. Lyu, "In ictu oculi: Exposing AI generated fake face videos by detecting eye blinking," *CoRR*, vol. abs/1806.02877, 2018.

[47] T. Jung, S. Kim, and K. Kim, "Deepvision: Deepfakes detection using human eye blinking pattern," *IEEE Access*, vol. 8, pp. 83144–83154, 2020.

[48] U. A. Ciftci, I. Demir, and L. Yin, "Fakecatcher: Detection of synthetic portrait videos using biological signals," *IEEE Transactions on Pattern Analysis and Machine Intelligence* (early access), pp. 1–1, 2020.

[49] G. de Haan and V. Jeanne, "Robust pulse rate from chrominance-based rppg," *IEEE Transactions on Biomedical Engineering*, vol. 60, no. 10, pp. 2878–2886, 2013.

[50] J. Hernandez-Ortega, R. Tolosana, J. Fierrez, and A. Morales, "Deepfakeson-phys: Deepfakes detection based on heart rate estimation," *CoRR*, vol. abs/2010.00400, 2020.

[51] L. Nataraj, T. M. Mohammed, B. S. Manjunath, S. Chandrasekaran, A. Flenner, J. H. Bappy, and A. K. Roy-Chowdhury, "Detecting GAN generated fake images using co-occurrence matrices," *CoRR*, vol. abs/1903.06836, 2019.

[52] M. Koopman, A. M. Rodriguez, and Z. Geradts, "Detection of deepfake video manipulation," in *The 20th Irish machine vision and image processing conference (IMVIP)*, pp. 133–136, 2018.

[53] C. de Weever, S. Wilczek, C. de Laat, and Z. Geradts. Deepfake detection through PRNU and logistic regression analyses. Technical report, University of Amsterdam, 2020.

[54] L. Guarnera, O. Giudice, and S. Battiato, "Fighting deepfake by exposing the convolutional traces on images," *IEEE Access*, vol. 8, pp. 165085–165098, 2020.

[55] Y. Qian, G. Yin, L. Sheng, Z. Chen, and J. Shao, "Thinking in frequency: Face forgery detection by mining frequency-aware clues," *CoRR*, vol. abs/2007.09355, 2020.

[56] C. Miao, Z. Tan, Q. Chu, N. Yu, and G. Guo, "Hierarchical frequency-assisted interactive networks for face manipulation detection," *IEEE Transactions on Information Forensics and Security*, vol. 17, pp. 3008–3021, 2022.

[57] K. Sun, H. Liu, Q. Ye, Y. Gao, J. Liu, L. Shao, and R. Ji, "Domain general face forgery detection by learning to weight," *Proceedings of the AAAI Conference on Artificial Intelligence*, vol. 35, pp. 2638–2646, May 2021.

[58] X. Wu, Z. Xie, Y. Gao, and Y. Xiao, "Sstnet: Detecting manipulated faces through spatial, steganalysis and temporal features," in *ICASSP 2020–2020 IEEE International Conference on Acoustics, Speech and Signal Processing (ICASSP)*, pp. 2952–2956, 2020.

[59] P. Korshunov and S. Marcel, "Deepfakes: a new threat to face recognition? assessment and detection," *CoRR*, vol. abs/1812.08685, 2018.

[60] B. Han, X. Han, H. Zhang, J. Li, and X. Cao, "Fighting fake news: Two stream network for deepfake detection via learnable srm," *IEEE Transactions on Biometrics, Behavior, and Identity Science*, vol. 3, no. 3, pp. 320–331, 2021.

[61] Y. Nirkin, L. Wolf, Y. Keller, and T. Hassner, "Deepfake detection based on discrepancies between faces and their context," *IEEE Transactions on Pattern Analysis and Machine Intelligence*, vol. 44, no. 10, pp. 6111–6121, 2022.

[62] F. Chollet, "Xception: Deep learning with depthwise separable convolutions," *CoRR*, vol. abs/1610.02357, 2016.

[63] Y. Yu, R. Ni, W. Li, and Y. Zhao, "Detection of AI-manipulated fake faces via mining generalized features," *ACM Transactions on Multimedia Computing, Communications, and Applications*, vol. 18, no. 4, pp. 1–23, Mar 2022.

[64] B. Liang, Z. Wang, B. Huang, Q. Zou, Q. Wang, and J. Liang, "Depth map guided triplet network for deepfake face detection," *Neural Networks*, vol. 159, pp. 34–42, 2023.

[65] Z. Yang, J. Liang, Y. Xu, X.-Y. Zhang, and R. He, "Masked relation learning for deepfake detection," *IEEE Transactions on Information Forensics and Security*, vol. 18, pp. 1696–1708, 2023.

[66] R. Han, X. Wang, N. Bai, Q. Wang, Z. Liu, and J. Xue, "Fcd-net: Learning to detect multiple types of homologous deepfake face images," *IEEE Transactions on Information Forensics and Security*, vol. 18, pp. 2653–2666, 2023.

[67] H. Chen, Y. Lin, B. Li, and S. Tan, "Learning features of intra-consistency and inter-diversity: Keys toward generalizable deepfake detection," *IEEE Transactions on Circuits and Systems for Video Technology*, vol. 33, no. 3, pp. 1468–1480, 2023.

[68] T. Wang, H. Cheng, K. P. Chow, and L. Nie, "Deep convolutional pooling transformer for deepfake detection," *ACM Transactions on Multimedia Computing, Communications, and Applications*, vol. 19, no. 6, pp. 1–20, May 2023.

[69] A. Vahdat and J. Kautz, "Nvae: a deep hierarchical variational autoencoder," in *Proceedings of the 34th International Conference on Neural Information Processing Systems*, NIPS '20, (Red Hook, NY), Curran Associates Inc., 2020.

Chapter 10

Empowering educators

Leveraging large language models for lecture preparation material development

Harshita Chandwani, Anuja Arora, Adwitiya Sinha, Neetu Sardana, and Deepika Varshney

10.1 INTRODUCTION

Large language models (LLM) have the potential to create or enhance existing solutions across various domains, and the current educational framework is one of them. In this domain, LLMs can be applied for tasks such as generation of quiz [1], planning of lecture [2], and suggestion for content [3]. In this same direction, there remains a notable gap in research concerning the generation of lecture preparation material specific to programming computer science courses. Literature suggests exploration of effective prompt engineering techniques to guide the language model specific to programming course-related lecture generation for neophyte coder.

Kojima et al. [4] mentioned in their work the success of LLMs is attributed to few-shot or zero-shot learning. Several studies have explored the use of large language models and attributed these models through prompt engineering using zero shot, few shot, conditioning few shots [5], instructional zero shot [6, 7], etc. The primarily used zero-shot learning works without any explicit training data and generate a generalize outcome of tasks. On the other hand, few shots use a very limited amount of data to achieve the desired outcome. The developer implements prompt tuning as well to generate task-specific data during training [8]. A significant concern in preparing lecture materials for coding courses is ensuring consistency. Literature recommends utilization of template prompts [9] to assist coding learners in grasping various algorithms and problem-solving concepts in a uniform and coherent manner. This research work is a notable effort towards generation of lecture preparation material using template-based prompting and aims to address the gap in various prompt engineering techniques by investigating the effectiveness of template-based prompting. The research objectives of this study are as follows:

- To generate lecture preparation material for programming/coding courses to serve as study material for professors.
- To explore the feasibility of template prompting for educational content generation

 DOI: 10.1201/9781003569619-10

- To evaluate performance of template prompting over conventional zero-shot and few-shot prompting for generating lecture preparation material by implementing a three-phased evaluation process.
- To explore utilization of Gemini pro for lecture preparation material quality assessment.

To achieve these objectives, comprehensive was developed involving the exploration of various queries, categorization of the queries, generation of lecture preparation material using various prompting techniques, and evaluation of the generated material using Gemini Pro which to our best of the knowledge has not been explored before. This research has made various contributions to the contemporary literature. First, by exploring the application of template-based prompting for generating lecture preparation material expanded the scope for use of large language models in education. Second, it renders educators and researchers with meaningful insights into the effectiveness of various prompting techniques and an understanding of the most suitable approach for generating high-quality lecture preparation material specifically for programming courses. Examining the quality and relevance of generated lecture preparation material using Gemini Pro demonstrates its capabilities as a content evaluator which is also an unexplored domain. Our work discusses about the literature studied, throwing light upon how the applications of LLMs are being explored in the education domain, the framework for generation of lecture preparation material, various prompting techniques explored, utilization of Gemini Pro with RAG for ranking, the experimental setup and results obtained from the three-phased evaluation in the following sections.

10.2 EXISTING LITERATURE SURVEY

The language models have already been used in education domain for diverse objectives such as to facilitate teachers by generating quizzes [1], defining lecture objectives [10], preparing lecture content [3, 11, and 12], providing suggestions based on the feedback of students [13], etc. Several research papers are studied in education domain specifically utilizing large language models and a few relevant research papers are summarized in this section.

One recently published most relevant research work by Spasić et al. [2] focuses on the development and testing of prompting techniques to enhance the generation of lecture plan for teaching programming to preschoolers. The authors employed three prompting techniques: (1) standard prompting, (2) standard prompting with role prompt, and (3) standard prompting with role, instructions, and seed words. The results demonstrate the capacity of the model to generate human-like responses aligned with the given task. However, there were notable differences among the generated lecture plans, particularly in terms of adaptability to preschoolers and adherence to unplugged programming principles. The findings suggest that carefully tailored prompts, incorporating role definitions and seed words, can significantly influence the quality and suitability of the generated lecture plans. Sridhar et al. [10] have used language model GPT-4 to automatically design learning objectives (LOs) for artificial intelligence courses. This study explores the efficiency of GPT-4 to produce high-quality LOs aligned with Bloom's Taxonomy. The result shows that GPT-4 can generate sensible and well aligned LOs with Bloom's taxonomy.

Another research work introduces an initiative to incorporate generative AI by applying large language model into online higher education classes at large public university in the USA. The implemented LLM extracts the rich metadata from lecture videos, segment lecture into coherent units, extract key definitions, examples, and procedural knowledge of taught topics. Even, questions to test student understanding of the taught topic using LLM are also generated. This research paper outlines several research directions to explore the full potential of LLMs for online education [13]. For quiz generation, EvalQuiz was introduced by Meißner et al. to automatically generate self-assessment quizzes in software engineering education using LLMs [1]. Traditional quiz creation is time-consuming process that require expertise and manual review of questions. EvalQuiz aims to alleviate this burden by leveraging LLMs, specifically GPT-4, to automate quiz generation based on lecture materials. The process involves several stages: topic modeling and keyword extraction, prompt engineering and question generation, assembling and validation of quizzes, and evaluation. Lecture material is processed to extract relevant keywords and generate prompts for the LLM. The generated quizzes are then assembled and validated, ensuring compliance with specified formats. Finally, the quality of the quizzes is evaluated, considering factors like relevance, correctness, and originality [1]. A personalized lecture summarization tool, HumSum, is developed by Kolagar et al. [14] for humanities students. HumSum collects user preferences through a scenario-driven survey to guide the implementation of the tool. Utilizing Streamlit for the user interface and map reduce function OF Langchain with GPT-4 model of OpenAI enabled effective summarization without collecting personal user data [14]. A survey published in a conference discusses the use of large language models specifically in the education domain [15]. It delves into the various stakeholders involved in education, including teachers, students, parents, educational institutions, government, and society, highlighting their roles and interactions within the educational ecosystem. Several applications of LLM in education domain are briefed in the manuscript and future implications of LLM in education are discussed to address the challenges in building a landscape where smart education becomes more personalized, diverse, and digitally enhanced [15]. Leiker and his group conducted experiments for the creation of course using an LLM under the guidance of prompt engineering and human oversight [11]. They meticulously examine both the advantages and hurdles associated with employing LLMs in the asynchronous creation of courses for adult education, training, and upskilling. Despite the immense potential of LLMs in accessing and synthesizing extensive human knowledge, reservations persist regarding their accuracy and reliability. Consequently, the study advocates for a balanced approach, suggesting the integration of human oversight to ensure the quality and clarity of the generated output [11]. Another research direction in this same domain is towards conversational tutoring system such as Ruffle&Riley [16]. This system automates scripting and orchestration for learning-by-teaching interactions and LLM agents engage users in free-form conversations. A user study comparing Ruffle&Riley to simpler QA chatbots and reading activities have shown no significant difference in post-test scores, but users rated Ruffle&Riley higher in understanding, remembering, helpfulness, and coherence [16]. Jauhiainen and Guerra research focuses on utilization of generative AI in primary school education. This theoretical study is done on 110 pupils of 8–14 years of age and studying in classes 4–6 in two different schools as participants in test lectures [12]. The results of the study demonstrated the feasibility of utilizing ChatGPT-3.5 to personalize learning materials to meet the knowledge and learning skills of pupils with varying levels of knowledge. Importantly, a clear majority of pupils reported enjoying the generative AI-modified material, indicating the potential for generative AI to support motivated learning and skills development in school education [12].

The studied research papers are usually used for theoretical subjects whereas lecture preparation material generation for professors using large language model for programming subject using a template-based prompting is still an unsolved and less explored issue.

10.3 LECTURE PREPARATION AND GENERATION PROCEDURE

The methodology involves using template-based prompt techniques [17] for generation of lecture preparation material using ChatGPT-3.5 for two important computer science fundamental subjects – Data Structure & Algorithm (DSA) and Object-oriented programming System (OOPS). The seven types of queries for which lecture preparation material could be generated and used by a professor were explored. Then, these queries were given as an input to ChatGPT-3.5 along with a fixed template respective to the type of query defining the structure of the material generated. Further from this pool of queries, 50 queries were extracted, and lecture preparation materials were generated for them using zero-shot prompting [18] and few-shot prompting [19] on ChatGPT-3.5. The materials generated using each type of prompt technique, for their respective query, were

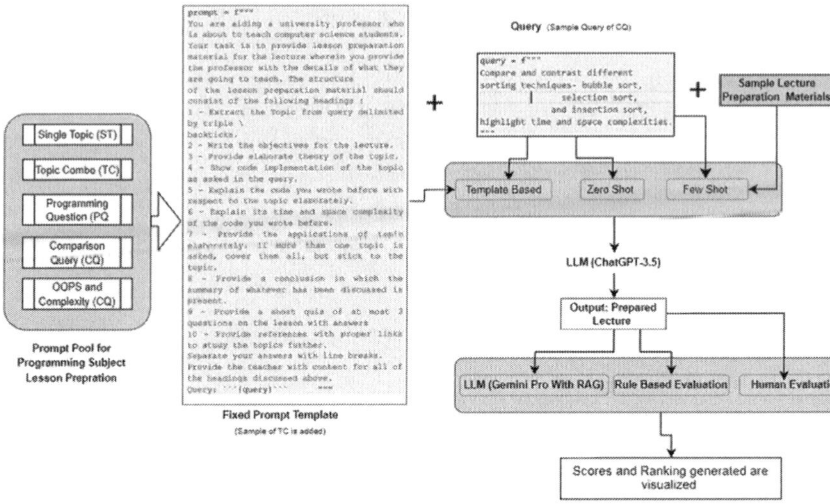

Figure 10.1 Illustration of our proposed lecture preparation process.

scored and ranked based on the structure and content quality using Gemini Pro and rule-based evaluation, respectively, followed by human evaluation. The framework of the proposed methodology has been visualized in Figure 10.1. In order to conduct this study, a set of 274 queries were explored in which DSA and OOPS topics taught in a classroom are included at least once. The queries were further categorized based on the type of content being asked and according to the type. A prompt structure was developed.

The considered five types of queries are as follows:

- **Single Topic Queries (ST):** Collection of 138 queries to provide lecture preparation material on a single topic such as Arrays, Linked Lists, Stack etc.
- **Topic Combination Queries (TC):** Collection of 70 queries to give lecture material on combination of topics, for example, explain a topic say A for topic B or explain a topic A for topic B in topic C or discuss topic A in topic B or illustrate topic A to topic B i.e. "give lecture preparation material on implementing stacks using linked list."
- **Comparison Queries (CQ):** Collection of 11 queries to arrange lecture material on comparison of various algorithm/techniques/methodology like prime query. For example, "Provide lecture preparation material comparing various sorting techniques like selection sort, bubble sort or insertion sort."

- **Programming Question Queries (PQ):** Collection of 48 queries to provide lecture material on a programming problem. For instance, "Share lecture preparation material on the programming problem fractional."
- **OOPs and Complexity Queries (OCQ):** Collection of seven queries to provide lecture material on OOPs and Complexity related topics. For example, "Share lecture preparation material on inheritance."

The lecture preparation material, for all the 274 queries, generated using template-based prompting were incorporated in a dataset called as the demonstration dataset.

10.4 PROMPT TECHNIQUES

There are several prompting techniques in large language models that are considered essential for successfully directing the model to generate the intended results. In our study, we have incorporated few shots, zero shots, and template-based prompting techniques.

10.4.1 Template-based prompting

Template-based prompting is a method used in natural language processing (NLP) where predefined templates or structures are provided to guide the generation of text. In the context of our study, template-based prompting has been used to craft different types of prompts tailored to different query types. The prompt consists of pointers defining the various headings a professor would expect in a lecture preparation material, thus guiding ChatGPT-3.5 in defining the structure of the lecture preparation material. The task of ChatGPT-3.5 is to fill in the content under these headings from its known knowledge. The various headings defined by the pointers in the prompt are as follows:

- **Topic:** The pointer extracts the topic from the query on which the lecture preparation material should be generated.
- **Objectives:** The pointer for this heading is to write objectives for the topic.
- **Theory:** The pointer provides an elaborate theory on the topic present in the query.
- **Code implementation:** Provides coding implementation of the topic and even though not specified in the prompt. All the codes are in python language in the output. For programming question queries, this heading asks for various ways a programming problem can be solved. For example, Knapsack problem can be solved using greedy algorithm and dynamic programming. The explanation and

complexity analysis of each approach is also asked, as shown in Figure 10.2.

- **Explanation of the code**: Provides proper explanation of the code written in code implementation section.
- **Complexity analysis**: this pointer details the time and space complexity of the code written in the code implementation section.
- **Beginner Programming Problem**: The pointer for this heading asks ChatGPT-3.5 to provide a beginner level programming question on the topic asked in the query. This exists only for single topic prompts.
- **The application of the topic**: The pointer for this heading asks ChatGPT-3.5 to provide application of the topic in real life.
- **Conclusion**: The pointer for this heading asks ChatGPT-3.5 to provide a conclusion of all that that has been discussed in the material above.
- **Short quiz**: The pointer for this heading asks ChatGPT-3.5 to provide a short quiz along with answers on the topic and what has been discussed in the lecture material.
- **References**: The pointer for this heading asks ChatGPT-3.5 to provide references to study the topic further. It is also mentioned to give links which are properly working.

Since there are different types of queries, the structure of prompts for the respective queries are also different, even though there are common headings in all types of prompts. The pointer differences in all the prompts are depicted in Figure 10.2.

10.4.2 Zero-shot prompting

In this prompting, ChatGPT-3.5 generates lecture preparation material for the given query without any additional illustrations. As shown in Figure 10.3, the zero-shot prompt [18] consist of the query enclosed within triple backticks and a brief instruction guiding the model to generate lecture preparation material for the mentioned. Zero-shot prompting, even though offers simplicity and flexibility, can lead to distinct responses depending on the interpretation of the model interpretation of the query and the complexity of the requested topic, which is not helpful in the problem area this study targets wherein a proper and consistent structure is required.

10.4.3 Few-shot prompting

In this prompting, ChatGPT-3.5 is provided with a few instances of queries and their respective lecture preparation material extracted from the demonstration dataset, referred to as "shots," to boost the accuracy and relevance of response. As shown in Figure 10.4, the few-shot prompt [19] consist of the query, natural language instructing the model to use the demonstrations

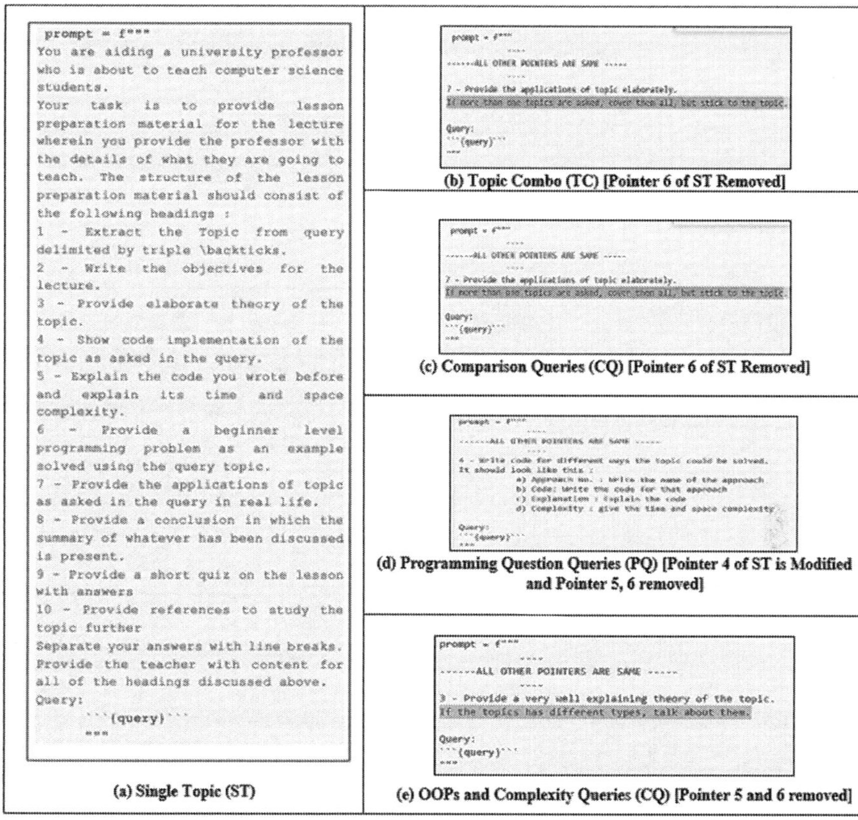

(a) Single Topic (ST)

(b) Topic Combo (TC) [Pointer 6 of ST Removed]

(c) Comparison Queries (CQ) [Pointer 6 of ST Removed]

(d) Programming Question Queries (PQ) [Pointer 4 of ST is Modified and Pointer 5, 6 removed]

(e) OOPs and Complexity Queries (CQ) [Pointer 5 and 6 removed]

Figure 10.2 Template for each query type.

to understand the structure and shots which may or may not be related to the topic, providing additional context and guidance for the model to generate a material on similar lines. Leveraging these instances, the model obtains a better understanding of what is expected and can generate more customized and thorough responses. The model learns from these examples, leading to more focused and informative outputs as compared to zero-shot prompting.

10.4.4 Gemini pro with RAG for ranking

Gemini Pro is an advanced natural language processing tool and, in this study, it has been given the task of determining the ranking of lecture preparation material generated using each prompting technique for each query on

```
ZS_prompt = f"""
      You will be given a query delimited by triple \
      backticks.You are job is to provide a lecture
      preparation material on the topic mention in the query.
      Query:
      ```{query}```
      """
```

(a)

Lecture Preparation Material:

Sorting is a fundamental operation in computer science that involves arranging elements in a specific order. There are various sorting techniques, each with its own advantages and disadvantages in terms of time and space complexity. In this lecture, we will compare and contrast three popular sorting techniques: bubble sort, selection sort, and insertion sort.

1. Bubble Sort:

- Bubble sort is a simple sorting algorithm that repeatedly steps through the list, compares adjacent elements, and swaps them if they are in the wrong order.
- Time Complexity: The worst-case time complexity of bubble sort is O(n^2), where n is the number of elements in the list. This is because in the worst case, the algorithm needs to make n passes through the list.
- Space Complexity: The space complexity of bubble sort is O(1) because it only requires a constant amount of extra space for temporary variables.

2. Selection Sort:

- Selection sort is another simple sorting algorithm that divides the input list into two parts: a sorted sublist and an unsorted sublist. It repeatedly selects the smallest element from the unsorted sublist and swaps it with the first unsorted element.
- Time Complexity: The time complexity of selection sort is O(n^2) in all cases, as it always makes n passes through the list.
- Space Complexity: The space complexity of selection sort is O(1) as it only requires a constant amount of extra space.

3. Insertion Sort:

- Insertion sort is a simple sorting algorithm that builds the final sorted list one element at a time. It takes each element from the unsorted part and inserts it into its correct position in the sorted part.
- Time Complexity: The time complexity of insertion sort is O(n^2) in the worst-case scenario, where the input list is in reverse order. However, it has an average-case time complexity of O(n) for partially sorted lists.
- Space Complexity: The space complexity of insertion sort is O(1) as it only requires a constant amount of extra space.

In conclusion, while bubble sort, selection sort, and insertion sort are all simple sorting techniques, they differ in their time and space complexities. Bubble sort and selection sort have a worst-case time complexity of O(n^2), while insertion sort has an average-case time complexity of O(n). Additionally, all three algorithms have a space complexity of O(1) as they require only a constant amount of extra space. It is important to consider these complexities when choosing a sorting technique for a specific problem.

(b)

Figure 10.3 Zero-shot prompt generated lecture.

the basis of the content quality using the context consisting of data related to the topic asked in the query retrieved using retrieval augmented generation (RAG). According to the Google, DeepMind report titled 'Gemini: A Family of Highly Capable Multimodal Models' [20], the pretraining dataset for Gemini included programming data which makes it a suitable choice for a content evaluator of computer science related material. Further it states that during training of Gemini models, a sequence length of 32,786 tokens were used, and it was observed that they effectively utilize their context length. The report also explored the performance of Gemini models for handling complex prompts, that is, prompts with multiple instructions and the per-instruction accuracy of Gemini Pro was 77.8 ± 2.0% making it adequate for the evaluation phase of this study as the prompt for Gemini Pro consists of the query and its respective lecture materials generated using each prompt techniques as discussed above, a context retrieved using RAG, which is explained in Section 10.4.4.1, natural language instructing the model to analyze and rank the content of lecture material generated by each prompting technique based on the context provided, resulting in a lengthy and a complex prompt as shown in Figure 10.5. The insights gained from implementing this technique is discussed in Section 10.5.
```

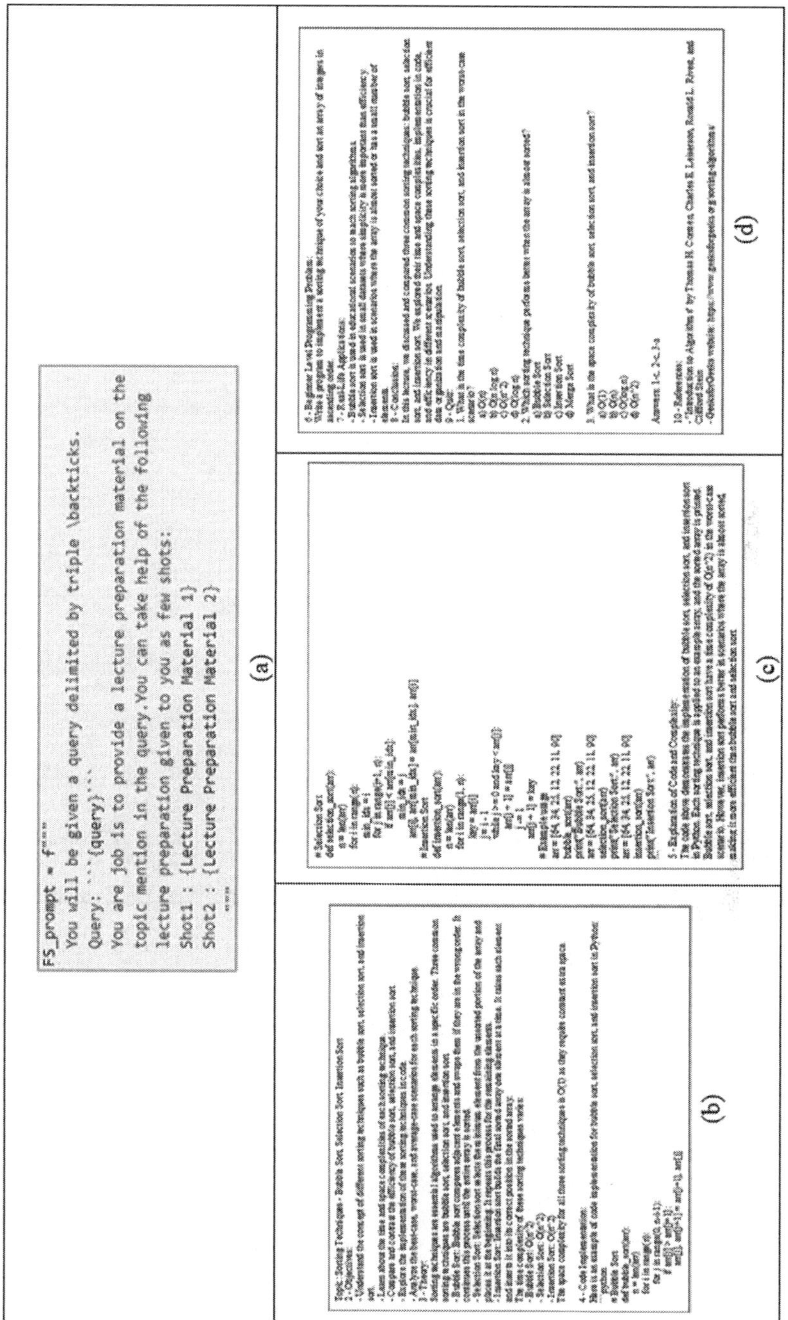

Figure 10.4 Few-shot prompting generated lecture.

```
Prompt = f""" You are a professor who teaches computer science to students.You need to rank lecture preparation materials for a query delimited by triple backticks.
The material is generated by 3 different prompt techniques called ZS,FS and TB.Ranking should be on the basis of the quality of the content under each heading .For
you to judge the quality, I am providing you with a context ,delimited by triple backticks, related to the query.Using this context, judge the materials and the material
which has the best content among  the three should be rank 1 ,followed by the next best material. If you think that 2 materials  similar in terms of content quality , give
them the same rank.The lecture material are :
'ZS': {zs}
'FS': {fs}
'TB': {tb}
Now the output should look like =  TB : the rank you gave to tb, FS: the rank you gave to fs, ZS: the rank you gave to zs
The ranking should not be similar to the order in which the materials are given to you. Analyze the material on the basis of
the criteria given using the context provided to you and rank them accordingly.
Query: ```{query}```    Context : ```{result}``` """
```

Figure 10.5 Prompt given to Gemini Pro.

10.4.4.1 Context retrieval using retrieval augmented generation

According to Ref [21], retrieval-augmented generation (RAG) refers to the retrieval of relevant information from external knowledge bases before answering questions with LLMs. RAG has been showcased to significantly enhance model performance for knowledge intensive task Since ranking lecture material is a knowledge intensive task, Gemini Pro is provided with context retrieved using RAG from a pdf containing topics present in DSA and OOPs coursework, stored in chunks in Chroma DB (a vector database). The context obtain has content related to the topic asked in query, and Gemini Pro is prompted to use this context to judge the lecture material in terms of context. This choice of exploring Gemini Pro with RAG as a content evaluator is backed by the attribution aspect of factuality of Gemini API models, that is, if grounded to a given context. The Gemini API models generate a response with the highest degree of faithfulness to the context as mentioned in [20].

10.5 EXPERIMENTAL SETUP AND RESULTS ANALYSIS

In this section, the experimental setup and outcomes of the study, which aimed to compare the effectiveness of template-based prompting (TB) with zero-shot prompting (ZS) and few-shot prompting (FS) in generating lecture preparation material for a diverse set of queries, is represented. Additionally, development and utilization of a front-end application was incorporated to facilitate the experimental process.

10.5.1 Front-end development

The front-end application uses React.js, providing an intuitive user interface for query submission and result visualization. The front-end allowed researchers to input queries of various types and select the prompting method for lecture generation, streamlining the experimental process. The screenshot of LPM system is show in Figure 10.6.

10.5.2 Test dataset composition and query distribution

Test set has been generated using demonstration data set (see Section 10.3). As discussed, the demonstration dataset consists of five query types: ST, TC, PQ, CQ, and OCQ. Test set is not chosen randomly to ensure a balanced representation of each query type. An equation-based approach is employed to determine the number of queries for each category as shown in Equation (10.1).

$$CQ * x + OCQ * x + TC * x + PQ * x = 50 \tag{10.1}$$

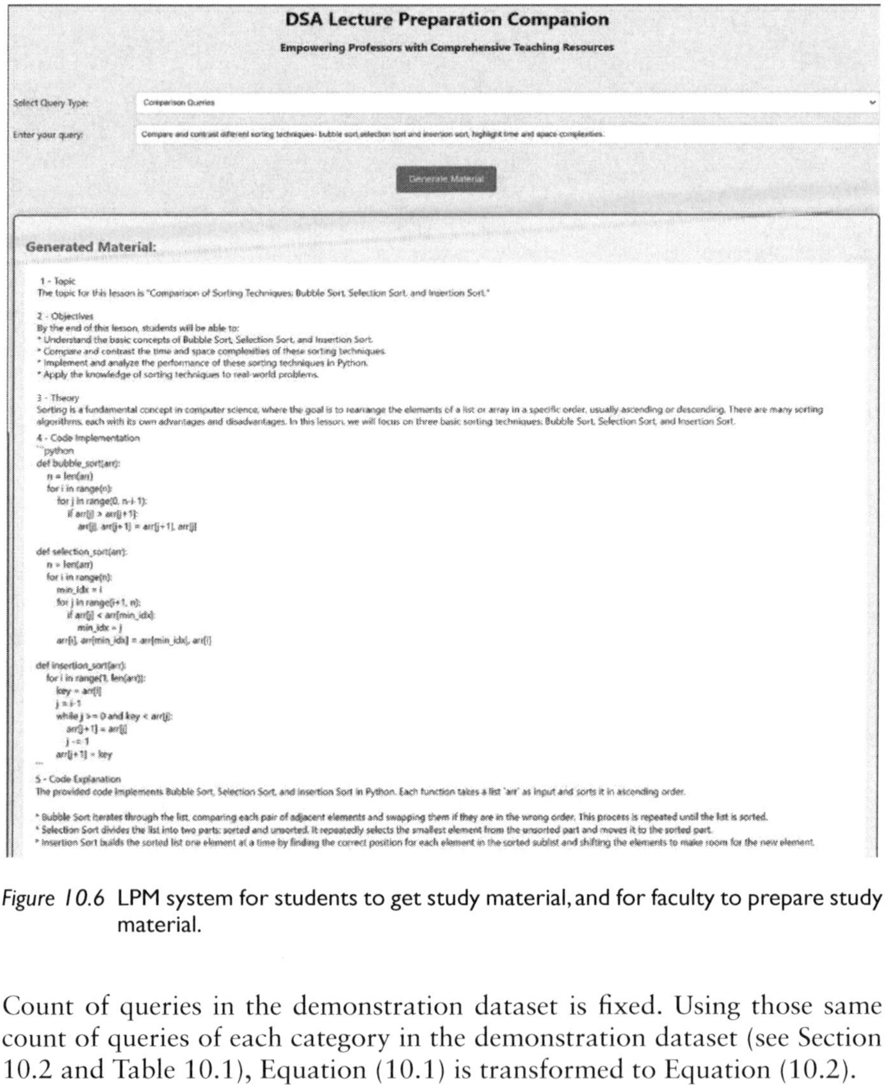

Figure 10.6 LPM system for students to get study material, and for faculty to prepare study material.

Count of queries in the demonstration dataset is fixed. Using those same count of queries of each category in the demonstration dataset (see Section 10.2 and Table 10.1), Equation (10.1) is transformed to Equation (10.2).

$$11x + 7x + 70x + 138x + 48x = 50 \tag{10.2}$$

In above Equation (10.2), x represents the scaling factor, and their respective constants represent the count of queries belonging to a particular query type in demonstration dataset. For example, in the term $11x$, 11 represents the count of comparison queries in the demonstration dataset.

Table 10.1 Dataset details for prompt technique ranking through Gemini

Query type	*# of queries in demonstration dataset*	*# of queries in test dataset*
CQ	11	2
OCQ	7	1
TC	70	13
ST	138	25
PQ	48	9

The test dataset of all the lecture material generated using prompt engineering techniques into consideration is combined into a list of dictionaries. This dictionary contains the query and the lecture material generated using zero shot (ZS), few shot (FS), and template-based (TB). These data are stored in a JSON file upon which evaluation is performed.

10.5.3 Evaluation

This section highlights the evaluation of our proposed technique through rule-based and human evaluation methods. Rule-based evaluation utilizes predefined rules and metrics to measure the performance of LLMs. Human evaluation relies on human judges to assess the quality of the LLM outputs. Combining both rule-based and human evaluation methods allows for a more comprehensive and precise assessment of LLM performance.

10.5.3.1 Rule-based evaluation

In this phase, the generated LPM are evaluated on the basis of a predefined structural criteria which focuses on finding specific keywords, within the generated content, like topic identification, objectives, theory explanation, code implementation, complexity analysis, applications, conclusion, quiz, and references which should be a header in an ideal LPM. For the presence of each component, the generated material is rewarded a score of 1 and conversely, the absence results in a score of 0. Finally, all scores are added and since there are total 10 headings, a score out of 10 is assigned to each lecture material. The results obtained depicts that for the 50 queries in test dataset, the score of LPM generated using template-based prompting (TB_score) ranges between 9 and 10, for the LPM generated using few-shot, the score (FS_score) ranges between 6 and 10, and for the LPM generated using zero shot (ZS_score), the score ranges between 0 and 5, as shown in Figure 10.7a. It can also be observed, in Figure 10.7b, that count of value of TB score and FS score as 10 is similar but TB score is not below 9 for any query which is the case in zero shot and few shot. Therefore, it can be

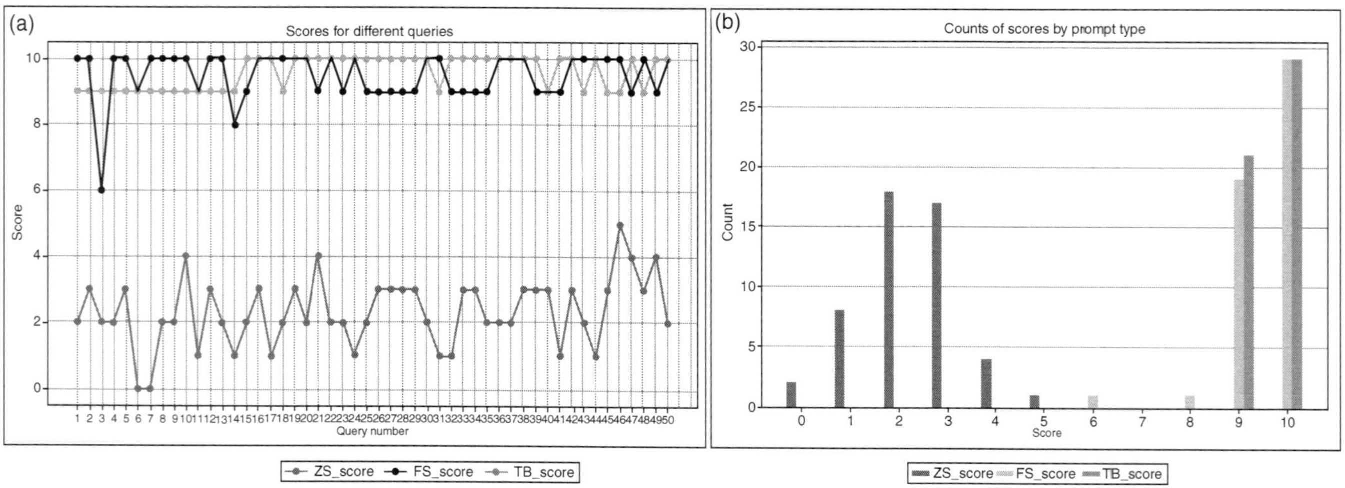

Figure 10.7 (a) Line graph representing the scores given to each of the prompt techniques. (b) Multibar graph visualizing the number of times a particular score is given to each of the prompt techniques for 50 queries on the basis of rule-based evaluation (structural criteria).

concluded that according to rule-based evaluation technique, TB prompt technique performed as good as FS prompt technique or better in terms of structural criteria.

10.5.3.2 Gemini Pro with RAG evaluation

In this phase, a prompt is developed for Gemini Pro as discussed in Section 4.4. Gemini Pro with RAG is used to rank the lectures based on content quality of LPM of test set. The ranked results were stored in a CSV file, with columns representing the query and the ranks assigned to ZS, FS, and TB lectures as presented in Figures 10.8 and 10.9. The left-side visualization in Figure 10.8 represents the number of times a particular type of prompt was ranked first. It is observed that out of 50 queries TB was ranked first maximum time. The right-side visualization in Figure 10.8 depicts the distributions of various ranks over each prompt technique that clearly depicts that TB was given rank 1 for most of the queries. For very few queries, it was ranked second but was ranked third way more times than it was ranked second. FS was majorly given second rank, and ZS was majorly given third rank. From the observation, it can be concluded that according to Gemini Pro, with respect to the context retrieved using RAG, even though a TB prompt technique performed better than other techniques for maximum queries in terms of content quality of LPM but for about 25% of the queries it was not the best prompting technique in terms of content. Figure 10.9 represents the ranking of TB for each query type by plotting the number of times TB has attained ranks 1, 2, and 3. It can be observed that for each query type, it has been ranked first more than it has been ranked second or third and for ST queries it has been ranked third for about eight queries.

The applied template-based prompt engineering result for test set are shown in Table 10.2. It can be observed in results that 48% times TB has received first rank, 18% times TB ranked second, and remaining 26% times TB ranked third in comparison to zero-shot and few-shot prompt engineering technique specifically for LPM generation.

10.5.3.3 Human evaluation

During this phase, from the test dataset a query of each type, which is total of five queries, along with their respective LPM generated using the three prompting techniques, that are, TB, FS, and ZS are shared with professors as a survey form. The survey required the participants to define their preferences as first, second, and third for the LPM generated using different techniques according with respect to the queries. The prompting techniques with which a particular LPM is generated were not revealed to the participants, and their ability to give similar preferences were enabled.

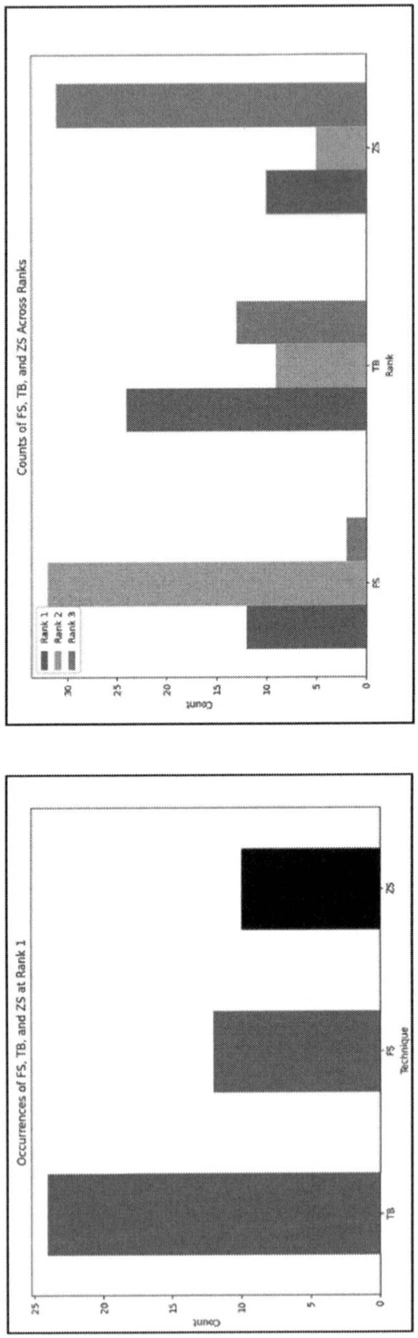

Figure 10.8 Generated lecture ranking comparison of template-based (TB), few shot (FS), and zero shot (ZS) using Gemini Pro With RAG.

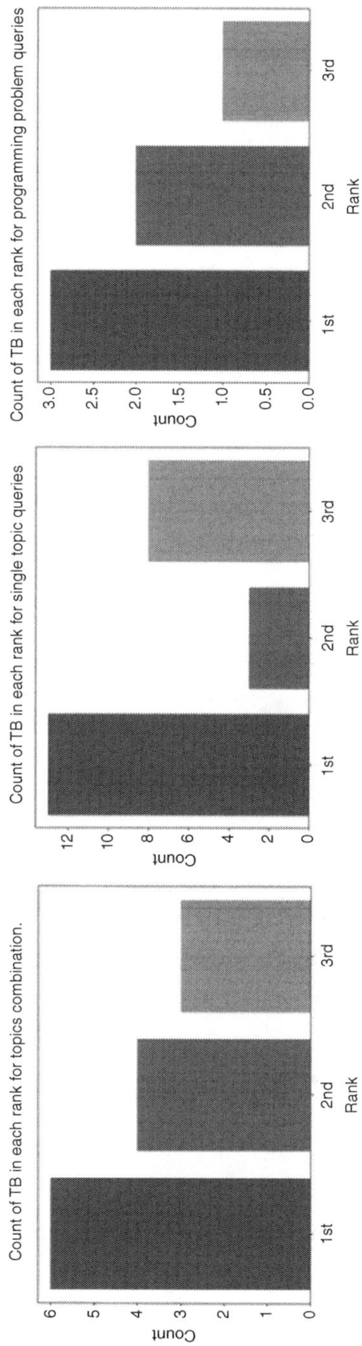

Figure 10.9 Template-based rankings order for different query type (a) TC, (b) ST, (c) PQ.

Table 10.2 Prepared lecture ranking results for template-based prompt engineering technique

Query type	*Total count in test dataset*	*Number of times TB was ranked first*	*Number of times TB was ranked second*	*Number of times TB was ranked third*
CQ	2	2	0	0
TC	13	6	4	3
ST	25	13	3	8
PQ	9	3	2	1
OCQ	1	0	0	1

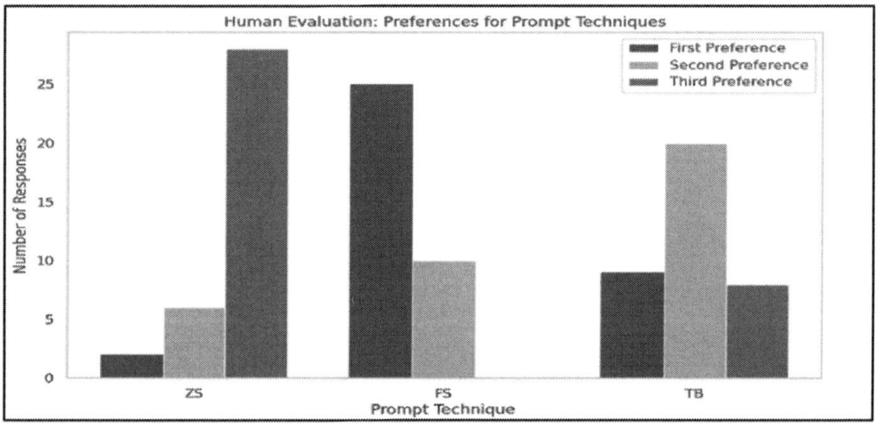

Figure 10.10 Multibar graph visualizing the count of all the prompt techniques as first, second, and third preference by a human evaluator.

The results obtained depicted that the LPM generated using FS prompt technique was majorly the first preference of the participants, followed by TB and then ZS. The LPM generated using TB was majorly a second preference, and the one generated using ZS was majorly given the third preference as visualized in Figure 10.10. This result also throws light on the gap between the sense of judgement of a human and LLMs like Gemini Pro as content evaluator.

10.6 CONCLUSION

This study involved leveraging advanced natural language processing techniques to generate education content. By employing various prompting techniques like template-based prompting, zero-shot prompting, and few-shot prompting techniques on Chat-GPT-3.5, lecture preparation materials were generated on topics related DSA and OOPs fundamentals and finally the performance of each prompting technique was compared by implementing a three phased evaluation. By carrying evaluation in three phases, we discerned the nuances in structure and content quality among the generated materials. The first two phases of evaluations, that are rule-based (structural criteria) where in for more than 55% of the queries in test dataset, TB has score of 10 out of 10 and for the remaining ones a score of 9 out of 10 which depicts structural completeness and Gemini Pro with RAG (content criteria) where in for about 48 % queries, the LPM generated using TB prompt technique are ranked first depicting content completeness, highlight the efficacy of template-based prompting in consistently producing well-structured and informative lecture content compared to zero-shot and few-shot prompting techniques for majority of the queries, but according to the third phase, that is human evaluation, the material generated using template-based prompting is majorly a second preference. Further, for evaluating generated education content, leveraging Gemini emphasizes on its value as a content evaluator as well as revealing areas where there is a scope of improvement to match human level evaluation. In conclusion, this study contributes to the ongoing discussion about the role of AI in education by throwing light on how content can be generated in educational systems by leveraging AI.

REFERENCES

[1] Meißner, N., Speth, S., Kieslinger, J., & Becker, S. (2024). EvalQuiz–LLM-based automated generation of self-assessment quizzes in software engineering education. In *Software Engineering im Unterricht der Hochschulen 2024* (pp. 53–64). Gesellschaft für Informatik eV.

[2] Spasić, A. J., & Janković, D. S. (2023, June). Using ChatGPT standard prompt engineering techniques in lecture preparation: role, instructions and seed-word prompts. In *2023 58th International Scientific Conference on Information, Communication and Energy Systems and Technologies (ICEST)* (pp. 47–50). IEEE.

[3] Bonner, E., Lege, R., & Frazier, E. (2023). Large language model-based artificial intelligence in the language classroom: practical ideas for teaching. *Teaching English with Technology*, 23(1), 23–41.

[4] Kojima, T., Gu, S. S., Reid, M., Matsuo, Y., & Iwasawa, Y. (2022). Large language models are zero-shot reasoners. *Advances in Neural Information Processing Systems*, 35, 22199–22213.

[5] An, Y., Xue, H., Zhao, X., & Zhang, L. (2021, August). Conditional self-supervised learning for few-shot classification. In *IJCAI* (Vol. 8, pp. 2140–2146).

[6] Lou, R., & Yin, W. (2024, March). Toward zero-shot instruction following. In *Proceedings of the 18th Conference of the European Chapter of the Association for Computational Linguistics: Student Research Workshop* (pp. 50–60).

[7] Wei, J., Bosma, M., Zhao, V. Y., Guu, K., Yu, A. W., Lester, B., ... & Le, Q. V. (2021). Finetuned language models are zero-shot learners. *arXiv preprint arXiv:2109.01652*

[8] Clavié, B., Ciceu, A., Naylor, F., Soulié, G., & Brightwell, T. (2023, June). Large language models in the workplace: A case study on prompt engineering for job type classification. In *International Conference on Applications of Natural Language to Information Systems* (pp. 3–17). Cham: Springer Nature Switzerland.

[9] Arawjo, I., Vaithilingam, P., Wattenberg, M., & Glassman, E. (2023, October). ChainForge: An open-source visual programming environment for prompt engineering. In *Adjunct Proceedings of the 36th Annual ACM Symposium on User Interface Software and Technology* (pp. 1–3).

[10] Sridhar, P., Doyle, A., Agarwal, A., Bogart, C., Savelka, J., & Sakr, M. (2023). Harnessing llms in curricular design: Using GPT-4 to support authoring of learning objectives. *arXiv preprint arXiv:2306.17459.*

[11] Leiker, D., Finnigan, S., Gyllen, A. R., & Cukurova, M. (2023). Prototyping the use of large language models (LLMs) for adult learning content creation at scale. *arXiv preprint arXiv:2306.01815*

[12] Jauhiainen, J. S., & Guerra, A. G. (2023). Generative AI and ChatGPT in school Children's education: Evidence from a school lecture. *Sustainability*, 15(18), 14025.

[13] Asthana, S., Arif, T., & Collins-Thompson, K. (2023) Field experiences and reflections on using LLMs to generate comprehensive lecture metadata. In *NeurIPS'23 Workshop on Generative AI for Education (GAIED)*, New Orleans, Louisiana, United States.

[14] Kolagar, Z., & Zarcone, A. (2024, March). HumSum: A personalized lecture summarization tool for humanities students using LLMs. In *Proceedings of the 1st Workshop on Personalization of Generative AI Systems (PERSONALIZE 2024)* (pp. 36–70).

[15] Gan, W., Qi, Z., Wu, J., & Lin, J. C. W. (2023, December). Large language models in education: Vision and opportunities. In *2023 IEEE International Conference on Big Data (BigData)* (pp. 4776–4785). IEEE.

[16] Schmucker, R., Xia, M., Azaria, A., & Mitchell, T. (2023). Ruffle&Riley: towards the automated induction of conversational tutoring systems. *arXiv preprint arXiv:2310.01420*

[17] Wu, T., Jiang, E., Donsbach, A., Gray, J., Molina, A., Terry, M., & Cai, C. J. (2022). PromptChainer: Chaining large language model prompts through visual programming. In *Proceedings of the 2022 CHI Conference on Human Factors in Computing Systems Extended Abstracts* (pp. 1–10). https://doi.org/10.1145/3491101.3519729

[18] Liu, C., Wang, C., Peng, Y., & Li, Z. (2024). ZVQAF: Zero-shot visual question answering with feedback from large language models. *Neurocomputing*, 580, 127505. https://doi.org/10.1016/j.neucom.2024.127505

[19] Gramopadhye, O., Nachane, S. S., Chanda, P., Ramakrishnan, G., Jadhav, K. S., Nandwani, Y., Raghu, D., & Joshi, S. (2024). Few-shot chain-of-thought driven reasoning to prompt LLMs for open-ended medical question answering. *arXiv preprint arXiv:2403.04890.*

[20] Gemini Team, Google Anil, R., Borgeaud, S., Alayrac, J.-B., Yu, J., Soricut, R., Schalkwyk, J. et al. (2023). Gemini: a family of highly capable multimodal models. *arXiv preprint arXiv:2312.11805* (2023).

[21] Gao, Y., Xiong, Y., Gao, X., Jia, K., Pan, J., Bi, Y., Dai, Y., Sun, J., & Wang, H. (2023, December 19). Retrieval-augmented generation for large language models: a survey. *arXiv preprint arXiv:2312.109972.*

Index